D0294747

PLANET
APE

Desmond Morris

with Steve Parker

MITCHELL BEAZLEY

PLANET APE

Desmond Morris
with Steve Parker

First published in Great Britain in 2009 by Mitchell Beazley,
an imprint of Octopus Publishing Group Ltd,
2–4 Heron Quays, London, E14 4JP

An Hachette Livre UK Company
www.hachettelivre.co.uk

Distributed in the USA and Canada by Octopus Books USA: c/o Hachette
Book Group USA, 237 Park Avenue, New York, NY 10017, USA
www.octopusbooksusa.com

A CIP catalogue record for this book is available from the British Library.

ISBN-13: 978 1 84533 4413

Art Direction **Pene Parker, Yasia Williams-Leedham**
Designer **Colin Goody**
Commissioning Editor **Peter Taylor**
Editor **Joanne Wilson**
Copy Editor **Daniel Gilpin**
Additional Text **Caroline Taggart, Daniel Gilpin**
Picture Research **Vickie Walters**
Production **Lucy Carter**

Set in Glypha
Colour reproduction by United Graphics
Printed and bound by Toppan, China

Gazing at the past
(Previous page) Sitting in front of an adult chimpanzee, a youngster takes in the scenic landscape of African tropical woodland, as great ape species have done for millions of years.

Watchful for the present
Great ape habitats, like the Democratic Republic of Congo forests of the bonobo ('pygmy chimpanzee') are fragmenting fast under a bewildering variety of human-induced pressures, from farming and logging for timber to mineral extraction.

Looking to the future
(Overleaf) In the midst of this destruction babies are still being born. But within the lifetime of this tiny female western lowland gorilla, the plight of the non-human great apes may go from critical to devoid of hope.

Contents

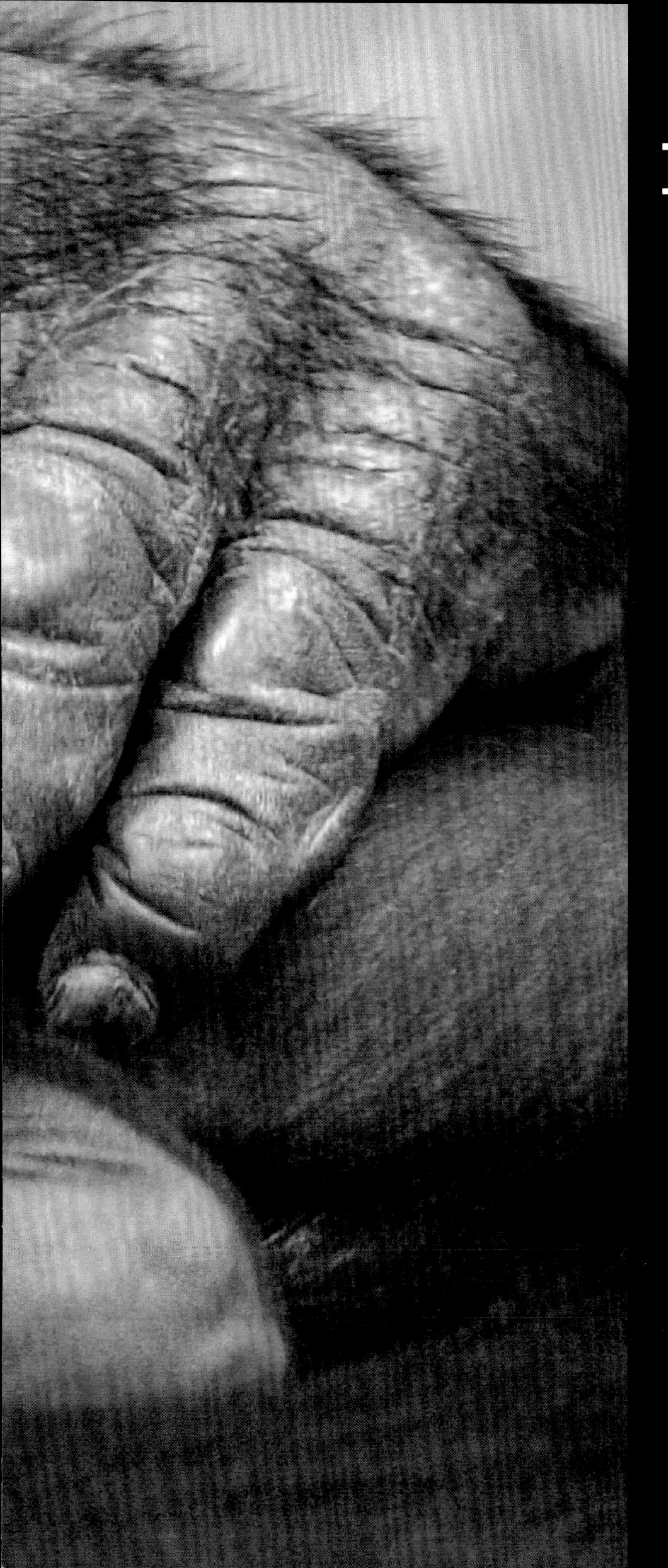

Foreword

When dawn breaks tomorrow on this small planet, seven types of great ape will get up out of their comfortable beds, stretch their limbs and start contemplating the first meal of the day. Six of these apes are on the verge of extinction; one is an unparalleled success. The six that are about to disappear are the western and eastern gorillas, the chimpanzee, and the bonobo, all of tropical Africa, and the two orangutan species of Borneo and Sumatra. The one that is on the verge of exterminating them is the naked ape, the animal usually referred to as the human being.

If this statement sounds melodramatic, consider the population figures. In the early years of the 21st century there are 6,600 million naked apes, occupying almost every corner of the Earth's landmasses. Forty years ago there were only 3,300 million of them, so their numbers have doubled since then. And they are still increasing. Current estimates predict a rise to 9,000 million by 2050.

Many species have ways of controlling their breeding rate, so that their populations stabilize and they do not outstrip their resources, but naked apes seem to lack these controls. Their numbers rise and rise and nothing is allowed to stand in their way. On a vast scale, wild places are tamed, forests become plantations, and grasslands become fields.

As a result of this human population explosion, each year there is less and less space for the other great apes and their present population figures confirm their increasingly desperate situation. At present it is estimated that there are probably fewer than 200,000 chimpanzees and bonobos, 140,000 gorillas and 40,000 orangutans alive in the wild. This makes a total of less than 400,000. To put it another way, for every hairy ape there are some 20,000 naked apes. So, in evolutionary terms, we are the winners and can congratulate ourselves on having almost obliterated our biological rivals. They may have retreated into the depths of the tropical forests and in this way escaped the attentions of our prehistoric ancestors, but it has failed to save them from our modern hordes: the loggers, the bushmeat hunters, the encroaching farmers and the rest.

If we do exterminate them, as now seems likely, it will be a sad day, a Pyrrhic victory, for two very special reasons. First, they are fascinating animals in their own right and, like all dramatically unusual species, enrich our lives on this wonderful little planet, where thanks to a host of astronomical coincidences, life forms have been able to develop and grow into more than 10 million distinct kinds. Of these, very few indeed have made the grade to become large, complex organisms, and we should cherish them, if only for their dramatic presence alongside us.

Second, and more importantly, we should respect our great ape relatives because they are a constant reminder to us that we are a part of nature and not above it. Genetically they are so close to us that their existence makes it impossible for any rational person to imagine that we humans have nothing to do with animal evolution. They force us to accept that we are part of the biosphere rather than being a separate, mystical creation.

Half a century ago, when scientists first really began to insist that we should view human beings as risen apes rather than as fallen angels there was an outcry. Many people were still clinging to the old idea that only humans have souls and that animals are brute beasts placed on Earth for us to use as we see fit. For centuries this arrogant philosophy had led to endless animal abuses and untold cruelties. From the mass animal slaughters of the Roman arena, to the torturing of animals as a royal entertainment in Elizabethan times, to the big-game hunting of the 19th century (which saw gorillas first killed in large numbers), few people gave a thought to the suffering of the creatures they were assaulting. If these animals were only brute beasts then obviously they had no feelings, other than rage and savagery when they were cornered and tormented.

But then, after Darwin introduced the concept of evolution, and animal-welfare societies began to flourish, attitudes slowly began to change. More and more people started to go out into the field to study natural history and to marvel at the wonders of nature. Instead of hunting, capturing, and killing wildlife, they watched it, and recorded what they saw.

At first, this was done mostly near to home. For a long time, we knew much more about British pond-life than we did about the exotic species that inhabited tropical forests. Africa was still a 'Dark Continent' full of mysteries and mortal dangers. In particular, the great apes, especially the gorillas, were looked upon as far too violent and dangerous to study at close quarters.

It was not until the middle of the 20th century that all this changed. Then, at last, observers began to penetrate the tropical forests and bring back first-hand accounts of great ape behaviour, and it was not what people were expecting. Gorillas turned out to be amazingly gentle giants and chimpanzees, although more volatile and irascible, were discovered to be inventive tool-users with a complex and subtly organized social life.

If explorers wanted to shoot wild animals now it was done with a camera rather than a gun. In particular, television documentaries opened everyone's eyes to the fascinating world occupied by our animal relatives. For the first time, voices were raised against the use of apes in medical research. And zoos were forced to provide their apes with social companions in large, complex enclosures. Solitary confinement in bare cages became a thing of the past in all reputable zoos, with the backward ones now looking increasingly primitive and out of step.

It was as though, without quite realizing it, the public was slowly accepting its close affinity with the great apes. Now, at last, people were beginning to worry about the survival of their primate relatives, rather than wanting to hunt them, experiment on them, or giggle at them dressed up as clowns. The hairy great apes had finally come of age in the consciousness of humankind. Attitudes of fear and ridicule were being replaced with respect and an intense curiosity about their lives. It is this curiosity that *Planet Ape* sets out to satisfy and in doing so will, I hope, also help in the struggle to save these magnificent animals before it is too late.

Desmond Morris

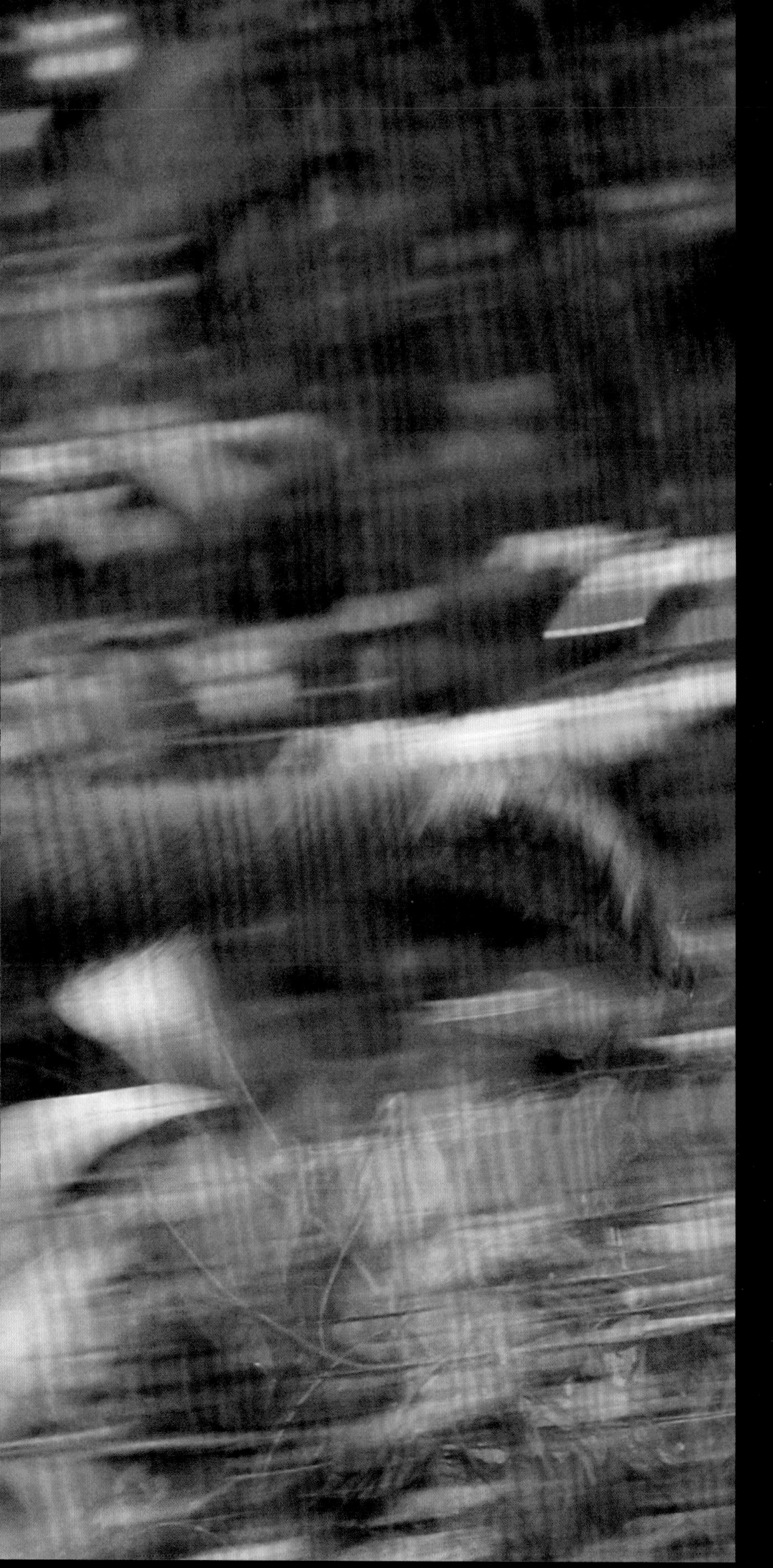

Great Apes

The living wild great apes – the orangutans, gorillas and chimpanzees – make a tight-knit family. They have shared traits that distinguish them from most monkeys and other primates, such as lack of a tail, special features in the skeleton, especially in the skull, an unusually large body and a big brain relative to the body. It is with the brain that great ape evolution really made progress, allowing the development of complex behaviour patterns and social structures.

What is a great ape?

How do they differ from monkeys?

Primates are mammals that took a special direction during the course of evolution, and that was to enlarge their brains and become the most intelligent animals on Earth. The primate group includes lemurs and bushbabies, tarsiers and pottos, monkeys and apes – and of course, humans. Physically their bodies remained rather primitive and all-purpose, but mentally they soared to new heights. Among other mammals, dogs, for example, have a much better sense of smell, cats have better night vision, bats have better hearing, cheetahs can run faster, and lions have more powerful jaws and claws. These mammals, like so many others, became specialists and superior to the primates in one way or another. But the primates were clever enough to outsmart them all and survive over a huge geographical range, in South America, Africa and Asia. And if you include the human primate, of course, the range becomes global.

Put very simply, there are three basic kinds of higher primates: monkeys, non-human apes and humans. The monkeys did not move very far along the primate evolutionary pathway towards better brains. Their brains are good, but not that good. They found a valuable niche for themselves in the treetops. There, with their lithe, athletic bodies and their grasping hands and feet, they could climb, leap and scamper about in search of nutritious food such as fruits, nuts, berries and insects. They could also employ their gymnastic skills to escape from the attentions of most predators. Some of them came down onto the ground to find extra food or to move from one group of trees to another, but when danger threatened it was an easy matter to leap back up into the safety of the high branches. Monkey life was a good life and they prospered and filled almost all of the warm forests on Earth.

Too brainy to be monkeys, but no match for humans, the apes pose an intriguing question.

At the other end of the scale, humans took the primate trend towards bigger brains relative to body size to the ultimate level, acquiring the ability to develop technological skills that would transform the planet. Somewhere between these two extremes, the chattering monkeys in the treetops and the chattering humans in their mega-cities, lie the non-human apes. Too brainy to be monkeys, but no match for humans, they occupy a sort of halfway house and pose an intriguing question. If they came this far along the primate path towards bigger brains, why did they not go further?

How did the non-human apes survive alongside humans?

Imagine an ancient ape ancestor, a kind of proto-ape from which all of today's great apes have descended. This must have been a type of monkey that came down from the trees and started moving about on the ground. In the process, it lost its tail. The tail had been useful as a balancing device when leaping and clambering through the trees, but now it was a nuisance and had to go. The proto-ape also had to grow bigger to defend itself from the larger predators that lived on the ground. And, perhaps most important of all, it had to become more cunning.

It was in this way that the large, tailless, big-brained ape came into being. Some of these ancestral animals started heading in the direction of modern humanity, their brains getting bigger and bigger. Along the way there were a number of evolutionary experiments, leading to several different kinds of proto-humans. Only one of these would go on to survive to the present day, our own species. With that in mind, one has to wonder how today's other great apes

Side branch
(Previous page)
Of the six non-human great apes the two species of orangutan are out on a limb in more ways than one. They are by far the most evolutionarily distinct, the most arboreal (tree-dwelling), least social and most geographically detached, compared to the four Africa-only species – two gorillas, chimpanzees and bonobos.

Top man
The gorilla – here a magnificently pensive and brooding western lowland male – is the summit of primate physical prowess.

survived. Why did they not disappear, along with all the other near-human animals?

The answer could be that today's non-human apes were species that took the step of retreating back into the forests from which we had all originally emerged. The Asiatic gibbons were the first to return. Still tailless and able to swing from their arms like all true apes, their bodies shrank until they were back at monkey size again, allowing them to take to the highest branches once more where no enemies could catch them. Later than the gibbons, the ancestral orangutans also fled from ground-level competition and predators in the Asian forests, heaving their heavy, tailless bodies up into the safety of the treetops.

In Africa, the ancestral gorillas disappeared into the dense, tropical rainforests near the equator and developed into giant herbivores, ground dwelling, but so powerful that they had little to fear. Last to go back were the chimpanzees and bonobos, already perilously close to humanity. Like the gorillas, they too retreated into the tropical African forests, away from primeval human eyes.

And there, in their forest retreats, all of the great apes were able to survive quietly until relatively recently. Dense rain forests were not a favourite haunt of human tribes, so the apes were largely left alone. In more recent times they have faced other pressures, being hunted to supply trophies for sportsmen, or captured alive to become exotic pets, zoo inmates, or subjects for laboratory experiments. As the human population has exploded, they have also begun to see their forest retreats chopped down, for timber and to make more space for agricultural land and human settlements. Happily, some of the forests that the apes inhabit are so unsuitable for human intrusion that they are still there. Now, at last, they also have a few friends, as well as enemies, from our own species – conservationists who do their best to defend apes and their habitats from destruction.

Dense rainforests were not a favourite haunt of human tribes, so the apes were largely left alone.

Why are apes so intelligent?

One of the features of the great apes that has puzzled some people is why they should have such big brains when their solution to survival was to regress to a form of existence not so very different from that of the much-smaller-brained monkeys. A typical monkey brain is about 100cc (cubic centimetres) in volume. By contrast, the brain of an orangutan measures about 350cc, that of a chimpanzee 400cc, that of a gorilla 500cc, and that of a human about 1,400cc. So although, in terms of brain capacity, a human is roughly three times as brainy as an ape, an ape is about four times as brainy as a monkey. Why do apes need all that brain power?

One answer often given to this question is that the apes need their big brains to cope with the ever-shifting social relations that exist within and between their groups, and with the constantly changing food supply, to which they must adapt their foraging behaviour. However, some people have argued that monkeys have similar problems and that they deal with them perfectly well. One field worker was so convinced that the wild chimpanzees he was studying were wasting their advanced brains, that he put forward a 'de-hominization' theory, in which he suggested that chimpanzees were much more like humans in the past. According to him, chimps were originally moving more and more towards the complex lifestyle of our own remote ancestors, but when they came into competition with them, they lost the struggle and had to beat a hasty retreat back into the darkest corners of the forests. There they still sit, living out a simple, monkey-like existence, with their brains not yet reduced back to monkey size.

Back to the forest
Bonobos (sometimes still called 'pygmy chimps') spend more time on the ground than chimpanzees

Evolution of the primates

In his book *The Descent of Man* (1871), which followed *On the Origin of Species* (1859), Charles Darwin wrote that 'man is descended from a hairy, tailed quadruped, probably arboreal in its habits'. That quadruped would perhaps have been not unlike one of the modern prosimians, the most primitive primates alive today.

The primate family tree shows the relatedness of different primate groups to one another and also the sequence in which they evolved. The shorter the journey back along the path back to the earliest primate group, the more ancient that group is. So we can see that the prosimians are more directly derived from the most ancient or 'basal' of the living primates and that humans separated from the African great apes more recently than did orangutans, for example.

The earliest known primates, such as *Plesiadapsis*, are known from fossils more than 50 million years old. Since then the entire order primates ('pry-mate-eez') has evolved into hundreds of species, both living and extinct. Apes evolved from the same ancestral line as the modern Old World monkeys. The earliest apes appeared on Earth some time between 25 and 20 million years ago. Later, our own group, the humans, evolved from a common ancestor we share with the modern bonobo and chimpanzee. The family tree shows this and it also shows that, for example, bonobos and chimpanzees are more closely related to us than either we or they are to gorillas.

GIBBONS

The first ancestral apes appeared around 22 million years ago. It is thought that they quickly split into the lesser apes (which today comprise the gibbons) and the great apes. Fossil evidence of the lesser apes, however, is extremely sparse.

OLD WORLD MONKEYS

The earliest fossils of Old World monkeys (Cercopithecidae) as we now know them date back more than 25 million years. It may be that this group appeared earlier, around the same time as New World monkeys. If so, their fossils have yet to be found.

NEW WORLD MONKEYS

The first monkeys are thought to have appeared perhaps 50 million years ago. From them appeared two lines, one carrying the ancestors of the New World monkeys (Platyrrhini) and the other the ancestors of the Old World monkeys and apes. The oldest fossils of true New World monkeys date back almost 30 million years.

PROSIMIANS AND COUSINS

Prosimians are the most primitive living primates – the least changed from the early primate-like mammals. The first prosimians appeared on Earth around 55 million years ago.

Relative values
As time went by, more and more different monkeys and apes evolved from our common primate ancestor (at centre). The further from the root of the family tree they moved, the more developed their brain became.

ORANGUTANS

Todays great apes are the last of much larger and more diverse lines. The ancestors of modern orangutans split away from the ancestors of African great apes (including humans) some time between 12 and 13 million years ago.

GORILLAS

Of all the African apes, gorillas are the least related to us. Their ancestral line is thought to have split away from the one carrying our ancestors and those of chimpanzees and bonobos around eight million years ago.

BONOBOS

Bonobos and chimpanzees are split geographically by the Congo river. They first began to evolve into separate species around 2.5 million years ago.

CHIMPANZEES

Together with bonobos, chimpanzees are our closest living relatives. Exactly when the human ancestral line split off from that of chimpanzees and bonobos is a subject of great debate. However, current research suggests that this happened around six million years ago.

PRIMATE ANCESTOR

The earliest known primate-like mammal is *Purgatorius*. About the size of a rat and probably insectivorous, it lived in what is now North America around 65 million years ago.

HUMANS

Our own species (*Homo sapiens*) first appeared on Earth less than a quarter of a million years ago. More primitive, ancestral humans, however, date back to the point at which our ancestral line split from that of chimpanzees and bonobos.

Where do the great apes live?

For planet ape, read planet human. The wild apes' closest living cousin, the human being *Homo sapiens*, has achieved a virtually global distribution and expanding populations in almost all parts of the world. In stark contrast, the non-human apes are suffering shrinking distributions, diminishing habitats and declining numbers.

WESTERN GORILLA

Gorilla gorilla

Distribution: Nigeria, Cameroon, D. R. Congo, Angola

Range: 400,000 million square km (154,500 million sq miles)

CHIMPANZEE

Pan troglodytes (approx. 4 subspecies)

Distribution: Western subsp. Guinea–Mali–Nigeria; Nigeria subsp. Nigeria–Cameroon; central subsp. Cameroon–CAR–DR Congo; eastern subsp. CAR–Sudan–Tanzania–Zambia.

Range: 2.3 million sq km (900,000 sq miles).

BONOBO

Pan paniscus

Distribution: D. R. Congo. They live south of the north-bending crook of the River Congo and its tributary the Lualaba, and are contained to the south by the Sankuru and Kasai river systems.

Range: 370,000 sq km (143,000 sq miles).

MOUNTAIN GORILLA (SUBSPECIES)

Gorilla beringei beringei and others

Distribution: Limited distribution around the D. R. Congo–Uganda–Rwanda borders.

Range: Virungas range, 375 sq km (145 sq miles) and Bwindi Impenetrable Forest, 210 sq km (81 sq miles).

In the wild, great apes are confined to the continents of Africa and Asia. The two orangutan species are limited to small regions of the large islands of Borneo and Sumatra, sharing some of their range with the lesser apes or hylobatids, including the gibbons and siamang. Orangutans prefer primary rainforest with relatively unbroken canopy, from lowland swamps to elevations of about 1,500 metres (4,900 feet).

> Non-human apes are suffering shrinking distributions, diminishing habitats and declining numbers

The other four non-human ape species are all found in West and Central Africa, and again their ranges overlap in some locations. The western gorilla inhabits mainly lowland equatorial forests, although it is known to visit baïs (forest clearings) along with elephants and other large mammals. Its range stretches from coastal West Africa to Central Africa, bounded to the east mainly by the Rivers Congo and Oubangoui, to the north by the River Sanga and the closed forest boundary, and to the south by the gradation from closed forest to drier scattered woodland savannah. Eastern lowland gorillas prefer similar vegetation but at higher altitudes, including swampy glades, bamboo forests and alpine zones between about 1,000 and 2,400 metres (3,300–7,900 feet) of elevation. The subspecies of the eastern gorilla known as mountain gorillas are restricted to mountainous and bamboo forests between about 2,800 and 4,000 metres (9,200–13,100 feet).

The chimpanzee is the most adaptable non-human ape in terms of habitat, ranging from dense rainforest to open wooded areas and scattered woodland-savannah. The bonobo is separated from the chimpanzee by the River Congo. Most of its habitat is primary rainforest between 300 and 750 metres (1,000–2,500 feet) in elevation, but towards the south of its range it lives in more scattered savannah woodland.

BORNEAN ORANGUTAN

Pongo pygmaeus

Distribution: More than 300 scattered populations in Borneo's south-central Kalimantan and north-west Sabah regions, and smaller, fragmentary populations in East Kalimantan and West Kalimantan–Sarawak.

Range: Less than 50,000 sq km (19,300 sq miles).

SUMATRAN ORANGUTAN

Pongo abelii

Distribution: A dozen or so remaining populations in Acer province (Nanggroe Acer Darussalam), from Guning Leuser southwards to Singkil Swamp, with further groups in West Batang Toru and East Sarulla.

Range: Less than 15,000 sq km (5,800 sq miles).

Isolated pockets

Looked at from a global perspective, the ranges occupied by the world's wild great apes are actually rather small. To say that apes are found in Africa and Asia is true but somewhat misleading. Africa and Asia are the world's two largest continents, but wild apes inhabit only a tiny fraction of each of them. In Africa they are largely confined to the Congo rainforest – a vast area of wilderness but a small part of the continent nonetheless. The areas occupied by Asia's wild great apes, the orangutans, are even smaller. They are restricted to small parts of two large islands in South-east Asia and do not occur on the mainland at all.

Prehistoric range

While today's wild great apes are concentrated into relatively small areas, it was not always that way. With the exception of our own species, great apes have always been confined to the Old World, but they once roamed over a much larger proportion of it than they do now. Extinct human species occurred in Africa, Asia, and Europe (see pages 66–67), and other great apes were similarly widespread. One of the first true great apes was *Dryopithecus fontani*, a bonobo-sized tree dweller that existed around 12 million years ago. It lived in the then subtropical forests of what is now France and was followed by several similar species that ranged right across Europe, from southern Spain to Georgia.

Around the same time as the *Dryopithecus* lineage was conquering Europe, another primitive genus of great apes, known as *Sivapithecus*, was living in Asia. Like the various species of *Dryopithecus*, these apes were tree dwellers, living and moving in a similar way to today's orangutans. *Sivapithecus* fossils have been found in Turkey and China, but the vast majority are known from Pakistan. These apes probably gave rise to other, later species and are, in fact, thought to be the direct ancestors of orangutans.

Lost giants

As well as the modern genus *Pongo* (the orangutans) another recent, but now extinct, genus of great apes may have evolved from *Sivapithecus*. They were giants, the largest apes ever known, and this is reflected in the name of their genus, *Gigantopithecus*. Their fossils are known from China, India, and Vietnam and date from around one million to just 300,000 years ago. Perhaps three species have been identified and all were enormous. The largest, *Gigantopithecus blacki,* is thought to have weighed up to 540kg (1,200lb) – twice as much as the largest male gorilla. Standing up on its hind legs it may have been as much as 3 metres (10 feet) tall. Like most of the non-human prehistoric great apes, *Gigantopithecus* are thought to have fed primarily, if not exclusively, on plants. In many places their fossils have been found alongside those of primitive pandas and this has led some scientists to suggest that they might have lived mainly on a diet of bamboo. Exactly why they went extinct is a mystery, but they may have been hunted by early humans. Fossils of one species that outlived them, *Homo erectus*, known for a time as *Homo pekinensis* or 'Peking Man', were found in eastern China in the 1920s. Widespread *Homo erectus* is known to have lived also in Java and other parts of East and South-east Asia, so it is possible that for a time it shared part of the giant apes' range. While *Dryopithecus* species inhabited Europe and *Sivapithecus* and their descendants lived in Asia, Africa had its own rich variety of non-human great apes. Fossils from various genera have been found in many parts of the continent where wild great apes no longer live. The fate of these extinct species, as for *Gigantopithecus*, may never be known for sure. It is possible that early humans hunted them, but they may have been victims of climate and habitat change. Certainly some places where their fossils have been found, such as Ethiopia, would not be able to support populations of wild great apes today.

With the exception of our own species, great apes have always been confined to the Old World, but they once roamed over a much larger proportion of it than they do now.

Tied to trees
(Previous page)
More than other great apes, orangutans depend on dense rainforest with continuous canopy. Here a youngster takes tentative 'steps' – swings – away from its ever-attentive mother.

Anything goes
Chimpanzees are the most adaptable of the apes – dry savannah woodland suits them just as well as dense rainforest incorporating mangroves like this.

Modern apes

Apart from our own species, the great apes that remain are fundamentally forest creatures. Chimpanzees may survive in wooded savannah, but never in habitats devoid of trees. Some 10,000 years ago, the estimated range of this great ape was almost three times its current area of distribution, extending north into what is now the dry Sahel south of the Sahara Desert (but which was then a much moister, partly forested landscape), north-west as far as present-day Mauritania, and south-east into central Tanzania. Likewise gorillas had a continuous distribution across West and Central Africa, joining what are now the separated western and eastern populations. Gorillas live in equatorial forests primarily because of their prodigious, vegetarian appetites: they would have trouble finding enough plant matter to fuel their bulky bodies outside of these lush habitats. Orangutans find their food up in the branches and much of their diet is fruit. Only tropical rainforests, where such food is available all year round, can support them. This reliance on forests has been the wild great apes' protection, keeping them largely hidden from our own species until relatively recently.

Family Portraits

There are six species and a number of subspecies of wild, non-human great apes. The six 'portraits' in this chapter focus on the most familiar and representative of each species, while including details of the others. The great apes' close cousins, those engaging 'little apes', the gibbons, and some colourful and intriguing representatives of Old and New World monkeys and prosimians appear at the end of the chapter.

Location:	Borneo (3 subspecies in NW, NE, and SW of island)
Habitat:	Tropical and subtropical forest
Population:	c.30,000 (declined over 50% during the last 60 years)
Status:	Endangered: possible extinction within a century

Bornean orangutan

It is easy to understand the way of life of wild gibbons, chimpanzees, and gorillas, but orangutans are something of a mystery. There is an old joke that says the gorilla started out as an ape that wanted to be a sumo wrestler, the chimp wanted to be a boxer, the gibbon wanted to be an Olympic gymnast, and the orangutan wanted to be a gibbon but couldn't make the weight. Behind this joke lies the fact that there is something decidedly odd about such a heavy-bodied ape, with such slow and deliberate movements, spending so much time in the trees. Other arboreal primates are typically lightweights, able to scamper about in the treetops, leap athletically from tree to tree, and clamber rapidly from one branch to another with breathtaking dexterity. The orangutan, in complete contrast, moves more like a ginger-haired sloth. What is the secret of the orangutan's puzzling lifestyle contradiction? To understand that we need to study the males.

Orangutans are the least sociable of the great apes. The males are solitary and are twice the size of the females. In the wild, a typical adult female weighs about 35–45kg (77–100lb) and an adult male over 70kg (150lb). Orangutans feed largely on a variety of fruits. If fruits are scarce, they will turn to other food sources such as leaves, ginger stems, bark, and small animals. Their rainforest environment allows them to enjoy a highly varied diet. In one study it was discovered that orangutans consumed no fewer than 400 different types of food.

Heavyweight champions

Big male orangutans, both Bornean and Sumatran, are always hostile to other males, defending the space around them with loud, bellowing roars, amplified by the use of a large, resonating throat sac. These vocalizations have a double function, because they also let the females know where the males are. The home ranges of each large adult male normally includes the home range of several females with young. When one of these females finally finishes rearing her offspring, when it is about seven years old, she will respond to the male's cries by approaching him for a mating.

The appearance of the adult male Bornean orangutan is extraordinary. Apart from his hefty body, he also has huge fatty flaps on the sides of his face that make him look even bigger and more threatening than he actually is. While male Sumatran orangutans have much less prominent cheek pads they too are enormous animals. Their size is directly linked to their success at holding a home range and hence breeding. This fact explains why male orangutans are as big as they are but not why they remain arboreal. The reason for that is three-fold. Firstly, their hands and feet are superbly adapted to climbing, but they are relatively clumsy on the ground. Secondly, the food they eat is mostly in the trees, and thirdly, until very recently, the forest floor was home to large numbers of

Look here
(Previous page)
In the tiny zoological corner that is the family of great apes, many humans identify with the chimpanzee's inventive cunning and up-front attitude.

Long time a baby
Apart from ourselves, no other species of animal cares for its offspring as long as the orangutans. And for them it is solely maternal attentiveness – not that of the males or other family members.

Big cheeks
The mature male Bornean orangutan has a squatter face, coarser hair, and more developed cheek pads or 'fatty flanges' than its Sumatran counterpart.

Almost too big
The orangutan is the world's largest tree-dweller. Females have greater freedom in the branches; mature males must watch their movements and test boughs for weight-bearing.

tigers, predators too big for even a large male orangutan to fend off. Significantly, we know from fossil evidence that 40,000 years ago orangutans were 30 per cent bigger than they are today. Those huge apes must have been forced by their own weight to come down to the ground where, presumably, they met their match and were killed off. What we see today is male orangutans with the most impressive physique that the trees they live in will still support. Were it not for their macho displays of physical power, orangutans might have become more athletically arboreal like the gymnastic gibbons. Instead we have this amazing lifestyle contradiction that gives us a heavyweight wrestler on a high trapeze, the largest arboreal animal in the world.

Quick and quick-witted

There is a second mystery concerning orangutans and that is why animals that live such a simple life should have such a high level of intelligence. Anyone who has worked closely with captive orangutans will vouch for the fact that, given the stimulus, an orangutan will show amazing mental abilities. Moreover, they are capable of tailoring their responses to fit the moment in a startlingly clever way.

To give an example, on one occasion I was about to show an adult female orangutan on a live television programme. The plan was to demonstrate her ability to open a series of boxes, arranged one inside the other, but she reached out to take the boxes before we were ready. As she did this, she was scolded by her keeper. Apes hate to be scolded, but this orangutan did not react immediately, as a chimpanzee would have done. Instead she waited until we were on the air. Then, without looking, she swung her long arm around in an arc, moving it backwards with great force, so that it struck the keeper in his solar plexus. He collapsed in pain while the orangutan continued to sit placidly awaiting the start of the test. It was a deliberate, calculated assault specially designed to fit the circumstances and, although I knew that orang over a long period of time and was even on one occasion allowed to watch her give birth, I never saw that particular action again. Rightly or wrongly, you get the feeling with orangutans that, even when they are at their most sedentary, they are quietly thinking things out.

Anyone who has worked closely with captive orangutans will vouch for the fact that, given the stimulus, an orangutan will show amazing mental abilities.

Another surprise, when working with captive orangutans, concerns their potential for speed. Orangutans have an acute fear of large snakes. On one occasion two London Zoo orangutans, which had spent their lives in captivity, were waiting to appear with me on a television programme. Without warning, a keeper carrying a giant python walked past them. Their immediate reaction, upon spotting this snake, was to flee in panic, climbing high up into the roof of the television studio and refusing to come down until the python had been removed. The speed with which they moved to escape was astonishing. Although normally rather slow and ponderous, orangutans are clearly capable of rapid movement when the need arises.

Falling numbers

Female orangutans only mate once their offspring can fend for themselves. As a result, these apes are extremely slow breeders and take a long time to recover from any population crash. The number of surviving Bornean orangutans is a matter of much debate, with estimates ranging from 15,000 to 69,000 individuals. Whichever figure is closer to the truth, the overall population is small and is rapidly getting smaller, as loggers move ever deeper into Borneo's forests. There is a real possibility that the Bornean orangutan could disappear from the wild completely and it is imperative that we learn as much as we can about them before it is too late.

Location: Sumatra (concentrated in north of the island)
Habitat: Tropical and subtropical forest
Population: c.7,000 (declined over 80% in the last 75 years)
Status: Critically endangered

Sumatran orangutan

The first time you come face to face with an adult male Sumatran orangutan is a moment that is hard to forget. Separated from its Bornean relative 1.5–1.7 million years ago, it has evolved a slightly different facial appearance and one that makes a strange impact. Its Bornean relative has a squarer head. The Sumatran head is taller, more diamond-shaped, and has much smaller cheek pads and vocal sacs. It also has a splendid beard with an elegance that the Bornean lacks. There is something about the configuration of its facial features that makes it difficult, during a close encounter, to resist the feeling that one is in the presence of a wise humanoid companion.

The Sumatran orangutan is very similar to its Bornean relative, so much so that until recently, when some minor chromosomal differences were established, the two were considered to be no more than subspecies. Indeed, many zoos have crossed Bornean and Sumatran orangutans and their offspring have proved to be fertile. If they ever met in the wild they could interbreed and act as a single species. This rather suggests that molecular biologists might be somewhat overstating their case. One recently announced: 'The Sumatran and Bornean orangutans have as many molecular differences as perfectly respectable species do. I feel we should treat them as different species.' As a result of such comments, male hybrid orangutans in certain zoos are said to have been given vasectomies to prevent them from breeding in captivity. However, the very fact that vasectomies were considered to be necessary does not help the argument that the two types of orangutan are distinct species, at least in the opinion of some scientists. This has led to an amusing controversy in which zoo authorities trying to keep the two ape lines separate from one another have been accused of racism and of fostering a policy of racial purity. The accusers view the molecular data as incomplete and misleading, and insist that variation within Bornean orangutans and also within Sumatran orangutans is as great as the differences between the two groups. The debate continues, but for the time being, despite the criticisms, the official position is that there are two distinct species.

Ape in danger

So, apart from its face, just how does the Sumatran orangutan differ from its Bornean counterpart? The short answer is very little. Its coat is sparser and generally lighter in colour. The typically dark red-brown of the Bornean orangutan is a paler cinnamon on the Sumatran. And the Sumatran's body is less stocky and squat. Although by no means sociable, the Sumatran orangutan is slightly less solitary than the Bornean. Beyond that, there is really very little to choose between them.

When it comes to numbers, the Sumatran orangutan is in even more danger of extinction than its Bornean relative. Towards the end of the 20th century it was estimated that there were about 12,000 Sumatran orangutans still surviving in the wild. By the time of the most recent survey, a few years into the 21st century, this figure had dropped to just 7,000. The prediction for the future is that the present population decline will continue at a rate of roughly 1,000 animals a year. If nothing is done to halt this decline, the Sumatran orangutan could become extinct in the wild in the very near future. Fortunately, there is a small population of these apes in zoos, where it looks as if the last remnants of the species may have to cling on if they are not to vanish altogether.

Drinking the rain The orangutan's exceptionally long interbirth interval, some seven to eight years, is a major factor in its plummeting population numbers.

A woman's work… Female orangutans spend most of their adult lives looking after their offspring, though in a lifetime of perhaps 40 years few will have more than three babies.

The main cause of the decline in the wild Sumatran orangutan population is logging. In recent years, vast areas of the rainforest these apes inhabit have been flattened to supply wood to the rest of the world. In the West, many believe that the Indonesian companies involved must secretly be ashamed of themselves, but when I asked one of their workers about their activities, on the spot where they were taking place, he waved a hand proudly towards the swathe of felled logs in front of him and told me how pleased they were with their progress.

When human populations start to explode, there is little chance of a gentle sharing of the land between humans and other animals.

Of course, chastising Indonesian loggers could be seen as hypocrisy. As some of these loggers point out, Europeans did exactly the same thing to their great forests centuries ago, destroying their wild populations of animals such as bears and wolves in the process. Furthermore, defenders of logging add, they are simply supplying the Western demand for wood. Who are we, then, to demand that the people of Sumatra should not advance their culture in the same way as we Westerners once did, creating more agricultural land and space for housing, while making financial gain from their natural resources? Who are conservationists, the locals might ask, other than foreigners who, having achieved their own industrial advancement, now want to keep places like Sumatra back in the dark ages, with wild beasts roaming at will and regions of the country off limits to humans?

A different perspective

The reason I have presented these anti-conservation arguments is that those of us living in the West so rarely consider the other side of the story. To a native of Sumatra, an orangutan is an agricultural pest, a source of meat, or perhaps, if a young one, a source of money if sold as an exotic pet. When human populations start to explode, as they are doing in Indonesia, there is little chance of a gentle sharing of the land between humans and other animals, of the kind one finds with small, remote hunter-gatherer tribes who can afford to retain their ancient respect for other species. Having made this uncomfortable point, the fact remains that the Sumatran orangutan is a gloriously odd and fascinating species and it would be a tragedy if it disappeared, simply because its very existence enriches our lives, even if we never manage to meet one in the flesh. Yes, there is a scientific value in having it alive for us to study more closely, but there is also a basic pleasure to be taken in sharing the planet with a close relative of ours, even if we learn nothing from it.

Saving the species

So how is the outcome of retaining the Sumatran orangutan as a living species to be achieved? Sadly, the only realistic way is by building up large captive-breeding colonies in various locations around the globe, where the climate is suitable for them and where they can be protected from all external dangers. If small regions of their natural forests can be protected, as game reserves or national parks, that too may be a way to keep them alive. Idealists who wish to see the whole of the natural orangutan range reserved for the apes themselves, unhindered by human encroachment, are very unlikely ever to see their ideals become reality. A more limited, more modest, and admittedly far less attractive goal may be the practical solution that will actually save the Sumatran orangutan from extinction.

Mother love
The relationship between mother and baby orangutan is perhaps the most intimate in the animal kingdom. A youngster may still occasionally be allowed to suckle when it is five years old, almost an adolescent.

A moment's rest
This young male Sumatran orangutan, resting on a liana, is safe enough, for today, in Bukit Lawang Sanctuary, within the World Heritage Site of Gunung Leuser National Park, Sumatra.

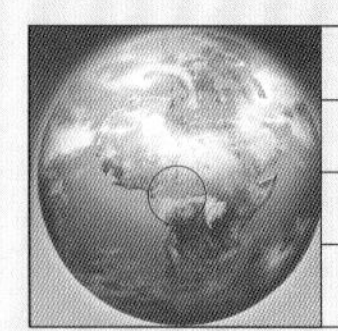

Location: Western central Africa – concentrated in Gabon
Habitat: Lowland tropical forests
Population: c.120,000
Status: Critically endangered

Western gorilla

Gorillas are the giants of the primate world, and in the popular imagination it is always the huge dominant male silverbacks that spring to mind whenever the word gorilla is mentioned. There is therefore something faintly ridiculous about the fact that the word gorilla comes from the ancient Greek word *gorillai*, meaning a tribe of hairy women.

Among the various primates, one often finds males that are bigger than females, but in the gorilla this trend finds its most extreme expression, the male being twice as heavy as any of his consorts. The western gorilla is sometimes called the crew-cut gorilla. It has short, bristly, greyish hair, compared with the long, dark, shaggy coat of its mountain-dwelling relatives. There has been a great deal of debate about exactly how many kinds of gorilla exist and that debate still continues today. The precise classification of animals has always caused controversy, dividing zoologists into two warring camps: lumpers and the splitters.

Defining species

The usual sequence of events goes like this: an intrepid explorer discovers an unknown animal in the wild and is so proud of the discovery that he or she labels it a new species. Someone else finds another, similar animal at a different location and also wants to be the discoverer of a new species, so that person too gives it a new name. These are the 'splitters'. This continues until there are many supposed species of a particular animal. At this point, the museum experts step in, compare the animals, decide that they are all local variants of the same species and reject the old names. These experts are the 'lumpers.' The next stage is new field-workers study the animals in the wild and, learning more about them, realize some of them are distinct species, so they split them up again. Finally, molecular biologists move in and examine the animals' DNA and more splits are made. This is what has happened to the gorilla. It all began in the 19th century, when a large number of gorilla 'species' were discovered. They included *Gorilla gorilla* (without beard), *G. diehli* (with beard and short hair), *G. beringeri* (with beard and very long, thick hair), *G. jacobi* (with short narrow rostrum), *G. castaneiceps* (with chestnut patch on head), and so on. There were 21 names listed altogether. But as more and more specimens were examined, it was discovered that the so-called species did not overlap in the wild, they simply occurred in different regions. Scientists concluded that they were therefore no more than local variants of the same species, which would interbreed freely if they ever met up. The new, single species of gorilla was given the scientific name *Gorilla gorilla*, and so the situation remained for most of the 20th century. In the last 20 years or so, however, new molecular evidence has given rise to a general feeling that the western and eastern populations should be viewed as different species, each with two subspecies. The western gorilla has the species name *Gorilla gorilla*. Its subspecies are the western lowland gorilla (*G. g. gorilla*) and the Cross River gorilla (*G. g. diehli*). The eastern gorilla is *Gorilla beringei*. Its subspecies are the mountain gorilla (*G. b. beringei*) and the eastern lowland gorilla (*G. b. graueri*).

Thoughtful time
Female lowland gorillas feel safe under the protection of the silverback. But they always remain alert to danger. Potential threats are monitored carefully before the group is warned.

Knowing stare
Western lowland gorillas take life fairly slowly and often pause to look around, sniff and listen for any unusual sights, scents and sounds.

Tall head
This male western lowland gorilla demonstrates the hugely powerful jaw and face muscles, exercised for many hours daily chewing bulky, low-nutrient food.

By the baï
West Africa's swampy forest clearings, known as baïs, are exploited by western lowland gorillas for juicy titbits – allowing researchers a clearer view of this wary, secretive great ape, as here at Moba Baï in Odzala National Park, Democratic Republic of Congo.

The western lowland gorilla

This subspecies once ranged across much of tropical West Africa, but today it exists in about 40 or 50 small, scattered pockets in Gabon, Angola, Democratic Republic of Congo, Central African Republic, Cameroon, and Equatorial Guinea. This is the gorilla that we know from captive specimens in zoos, but until recently it has been the least studied in the wild. The reason for this is that it has been so badly persecuted in the past that as soon as it sees humans approaching it fears the worst and flees in panic. As a result, field-workers have a hard time getting close enough to make detailed observations. Surprisingly, one of the best habitats for field studies has turned out to be swamp forests, where gorillas, normally terrified of water, will wade into the swamps to obtain the lush aquatic plants that grow there.

Away from the swamps, western lowland gorillas find it hard to locate high-quality food, and much of their waking day is spent in picking and eating much poorer quality plant matter. In this respect they are primates that have come to live like ungulates (hoofed mammals). Like most ungulates, through evolution gorillas have grown greatly in size, and now they are large and strong enough to defend themselves against virtually any predator. The one exception is our own species. Human hunters began slaughtering these magnificent apes mercilessly as soon as they possessed the necessary weapons to do so. Early paintings showing a huge male gorilla rearing up and savagely attacking a man who was holding a spear or a gun were images that terrified those who saw them. In reality, however, these pictures were depicting not an unprovoked onslaught by a marauding monster, but instead a pathetic, last-ditch attempt on the part of a dominant silverback to protect his females and young from slaughter. The shy, retiring, restrained herbivore that is the true gorilla was unforgivably portrayed as a violent assailant.

Changing attitudes

It was the thrill of seeing this dreaded monster conquered and confined behind sturdy iron bars that drew a fascinated public to zoos in the early days, to gawp at and tease the beast. For a long time zoos did not even consider housing gorillas in natural social groups or stimulating enclosures. If they could get one large male gorilla sitting forlornly in a cage, that was enough to draw the crowds, so that was what they did. Fortunately, in the second half of the 20th century everything changed. As we learned about the lifestyle of wild gorillas, it became clear that captive animals must be kept in much larger enclosures and in proper breeding groups. In many places zoos are at last breeding western lowland gorillas, rather than taking more from the wild. The fact that gorillas are breeding shows the success of the efforts made to house them in conditions with which they feel comfortable.

Location:	Rwanda, Uganda, Democratic Republic of Congo
Habitat:	Montane rainforest (2,225–4,267m/7,300–14,000ft altitude)
Population:	c.650
Status:	Critically endangered

Mountain gorilla

Most of our knowledge of wild gorillas comes from field studies, not of the common lowland forms, but of the extremely rare, almost extinct mountain gorilla, a subspecies of the eastern gorilla. The mountain gorilla is the shaggy-coated ape we now know from the studies of George Schaller and Dian Fossey, and from the famous television film of David Attenborough. Today, there are perhaps as few as 600 of them left alive in the wild and there are none in zoos.

The last mountain gorillas acquired by London Zoo arrived in the early 1960s. They were magnificent young animals, and it was hoped to breed from them and start a captive colony, but they succumbed to human diseases and were soon dead. They were the last to be seen outside their homeland where, it is not exaggerating to say, their subspecies' survival hangs by a thread. It has been estimated that 5,000 western lowland gorillas were wiped out a few years ago by an epidemic of ebola fever, and it is easy to imagine what such a virus would do if it arrived in the home of the last 600 mountain gorillas. Fortunately, the mountain gorillas are separated from their western cousins by a distance of more than 1,000km (620 miles), so for the moment they are safe from this particular threat, but increased tourism in their mountain retreats is bringing almost daily risks of the spread of human diseases. Those people who are so fascinated by the story of the mountain gorillas that they must get as close to them as possible, to experience a romantic oneness with nature, may turn out to be the very people that bring about their destruction. It may take no more than a single sneeze, bringing in a chest infection against which the animals have no protection.

Wealth of knowledge

It is to be hoped that this disaster will never happen, but if it does we will have a treasure trove of information about these unique apes by which to remember them. The studies undertaken to date have been among the most thorough of any animals in the wild.

Mountain gorillas live high up in the cloud forests at the point where Uganda, Rwanda, and the Democratic Republic of Congo meet, in the very centre of Africa. Nine-tenths of their diet consists of leaves, flowers, shoots, stems, and roots, which may be taken from as many as 142 different plant species. They supplement this occasionally with bark, fruits, soil (for minerals), and ants. A big male will consume 34kg (75lb) of vegetation every day. Because their diet is of such low quality nutritionally, mountain gorillas must spent at least half of the daylight hours eating. The other half is spent resting, and at night members of a group sleep close to one another in separate beds made freshly each evening. To be blunt, this is a bovine existence for a primate with such a big brain, a picnic party with no beginning and no end. Social excitements are few and far between. Personal relationships within a mountain gorilla group are relaxed and easy-going. When one group encounters another there are rarely territorial battles. Instead the

Family time
After decades of intensive observation, some mountain gorillas have become 'habituated' or used to human presence. This group is in Mgahinga National Park, Uganda, part of the Virunga Mountains region, where visitors can go 'gorilla tracking' in the hope of a close encounter.

Furry coat
Mountain gorillas are identified at once by their long hair, which is vital in their often cold and sopping-wet environment.

two groups usually do their best to avoid one another. Apart from humans, there are virtually no predators for these apes to worry about. Competition with chimpanzees is negligible, because the chimpanzees that share their range are essentially arboreal fruit-eaters and have little interest in the ground plants taken by gorillas.

The typical mountain gorilla social group consists of eight to ten individuals – one dominant male, three or four adult females and their young of varying ages. In up to 40 per cent of groups the dominant male does, however, allow a few other, lower-status adult males to join his party, where they can play the useful role of look-outs. Allowing these additional males into the group does have another advantage because, when the old silverback eventually dies, one of these other males can quietly take over. If no other adult males are present, there tends to be a complete break-up of the family group when the dominant male dies. This multi-male grouping is something that a western lowland gorilla would avoid at all costs, and seems to be one of the key behavioural differences between the eastern and western forms.

Although the basic social unit is a harem consisting of one huge male with several much smaller females, the male cannot rely on brute strength to keep his harem around him.

Gentle giants

Relations between male and female gorillas are generally friendly. Although the basic social unit is a harem consisting of one huge male with several much smaller females, the male cannot rely on brute strength to keep his harem around him. It would be far too easy in the dense forest environment these apes inhabit for a female simply to slip away and find another group if she were badly treated. So, despite their immense strength, males have no choice but to make their dominant role a benign one. It has also been observed that they are especially gentle towards their infants. Of all the great apes, gorillas seem to be the ones that are the least interested in sex. When they do copulate it is brief and far less impressive than one might imagine. Mating acts in captivity have been timed at 10 to 15 seconds, with an exceptional case lasting up to three minutes. During these observed copulations the males only made a few pelvic thrusts.

The first observations of wild matings told a slightly different story, suggesting that captive gorillas were perhaps inhibited by their artificial surroundings. One case involved 14 pelvic thrusts, and another, with the mounting of the female by the male lasting almost 5 minutes, no fewer than 300. The most surprising feature of these two matings, however, is that they both involved not the dominant male but one of his subordinates. Throughout both matings, the dominant male reclined peacefully on his resting knoll, no more than 6m away, and could see what was taking place. At the very least, one would expect him to leap up and come charging across, roaring and beating his chest, to separate the couple and chastise the subordinate male but, instead, he quietly permitted the mating to take place. For a well-fed silverback it would seem that a siesta takes priority over sex.

Later, more detailed field observations of mountain gorilla groups containing more than one male confirmed that dominant males do sometimes allow subordinate males to mate, but that the majority of matings are by the alpha male himself. Eleven of the fourteen females observed in this study mated with more than one male, revealing a level of male competition far below what one might expect. Far from being the brute of popular fiction, therefore, the gorilla, despite its great strength, is clearly the gentlest, friendliest, and most restrained of all the great apes – a fact that somehow makes our relentless extermination of these magnificent animals all the more tragic.

Young and old
A wide-eyed young mountain gorilla watches the scene intently, finding everything so interesting yet puzzling. In the background his enormous father has a restrained but watchful remit.

Highlight of the day
Mountain gorilla routine includes a great amount of sitting around, quietly picking and eating, followed by a pause, then more picking and eating.

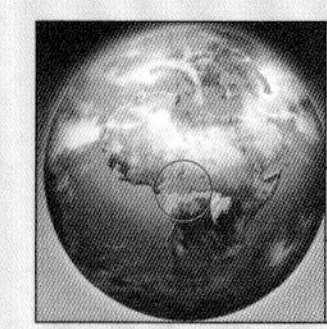

Location: Central and western Africa
Habitat: Deciduous rainforests, also grasslands and woodlands
Population: 100,000–200,000
Status: Endangered

Chimpanzee

The chimpanzee and the bonobo are considered our nearest living relatives, and the chimpanzee is the ape that shares most closely the great energy, curiosity, and social complexity of the human species. In the wild, each chimpanzee is subjected to a constantly shifting set of social relationships that require considerable intelligence, cunning, and flexibility to manage.

The first thing one notices about a chimpanzee is how expressive it is, compared with an orangutan or a gorilla. Chimpanzees have a wide range of facial expressions and vocalizations that cover all of their basic moods. There is a special play face in which the animal exposes its lower teeth but covers its upper ones. If a chimpanzee attacks another while making this face, the other knows that this is only play fighting. If a chimpanzee is tickled during the rough and tumble of play, it will utter a panting noise that is the equivalent of human laughter. Chimpanzees also have a hooting, long-distance greeting and a softer, close-quarter greeting. They have a food call, and a greeting-the-food call, formed from a combination of the food and greeting calls. They whine in distress and scream to express anger or fear. They drum their feet rhythmically against the buttress roots of large trees, possibly to signify social status, and they can perform several hand gestures – friendly, imploring, begging, or to express agitation.

Complex communication

The behaviours listed above are just some of the ways in which chimpanzees can communicate with their companions. This ability puts them into a different league from the taciturn gorilla or the introverted orangutan. The chimpanzee's sociability and extrovert volatility make it the ideal animal to study, and recent investigations, both with large captive colonies housed in huge enclosures and with wild communities in Africa, have time and again startled human observers. We now know that adult chimpanzees employ subtle political strategies to gain social standing and that they not only use tools but also make them and employ them in a variety of different ways. We have discovered that they will eat meat and will engage in cooperative hunting in which there is a division of labour, with some apes acting as drivers and others as ambushers, when cornering and killing colobus monkeys. We have even learned that chimpanzees may sharpen sticks to act as spears, with which they stab sleeping bushbabies in their nest holes. These and a host of other recent observations have confirmed what some scientists have long believed – that the chimpanzee is truly the most advanced animal on this planet, with the exception of our own species.

Sociable ape
Chimpanzees travel, feed, rest, groom, and bicker in small ever-changing clusters; the only truly enduring bond is between mother and offspring.

Changeable moods
This female chimpanzee's pensive expression could give way in a split second to raucous indignation or highly charged aggression, so rapidly does their disposition alter.

Interacting with humans

The best way to understand the subtlety and complexity of chimpanzee social life is to watch the way it tries to influence relationships within its social group. I was able to do this over a three-year period with a young male chimpanzee called Congo and was astonished at what

I saw. Congo appeared with me each week on a live television programme, the purpose of which was not to present the ape as a clown, but to demonstrate to a wide audience its extraordinary level of intelligence and inventiveness. While this was being done it was also possible for me to study the way in which the ape handled its social relations.

On one occasion Congo suffered a mild chest infection and had to be rested. In case this happened again, a second young chimpanzee was acquired and added to our human/ape family. Congo greeted the new arrival with interest and initially there was no sign of trouble. But once Congo had recovered, the time came for the two chimpanzees to go on television together. With the performance about to begin, my assistant held Congo and the other chimpanzee ready, one in the crook of each arm. When Congo heard me mention his name and knew that he was about to appear, he gently leant across to his new companion, as if to kiss him. But then, at the last minute, he sank his teeth into the friendly, pouted lips of the other ape. Immediately all was chaos. There was blood and screaming, and the newcomer had to be rushed from the television studio. Congo triumphantly appeared on the screen, as he had always done, as the sole centre of attention.

Congo knew that each week, when he appeared on television, he was for a while going to be treated as the most important member of his human family.

This tale of Machiavellian behaviour may sound exaggerated, but not to anyone who has studied chimpanzees in the wild. Congo knew that each week, when he appeared on television, he was for a while going to be treated as the most important member of his human family. He sensed this in the way he became the focus of everyone's interest. This was a social role that meant a great deal to him, especially as he was a very young male and was, in a sense, being elevated above his true social status. He was not about to share that elevated position with another young ape.

Violence within the group

On another occasion, Congo managed to defeat a new human member of his social group in a similar way. The arrival of a new secretary disrupted what he considered to be a fixed set of social relations and he tried to attack her. She was told never to be alone with him but one day forgot this warning and entered his quarters, where he was sitting on a table, waiting to be fed. At that moment the food was being fetched and nobody else was present. Without hesitation, Congo leapt at the secretary and, with a savage bite, tore a large flap of flesh from her shoulder. He then did something that he had never done before. He jumped inside his sleeping cage and shut the door. It was always a struggle to persuade him to enter this cage at the end of the day, but on this occasion he leapt in of his own volition. When my assistant returned with the food, helped the injured woman and then turned to Congo, the ape's gaze was firmly fixed on the ceiling. On being scolded, he ignored the angry words that would normally have made him whimper and whine, or explode in a temper tantrum. Instead, he continued staring up at the ceiling with an expressionless face.

Many examples of this kind of hierarchical social behaviour could be given, but these two alone reveal the extraordinary complexity of the chimpanzee's social life. Translated into a wild context, where each animal in a group, as it grows older, will try out new ways of changing its social status, it is easy to see why chimpanzees are constantly alert. Finding food may be easy, making a bed each night may be straightforward, avoiding predators may not be that difficult, but keeping one's social standing in a healthy condition, day by day, is a huge challenge for any chimpanzee.

Trials of growing up Young male chimpanzees face a tough time breaking into the elite 'gang' of top males within the group. To begin with, they watch from a distance and work out alliances and rivalries.

Group politics The chimpanzee social hierarchy is complex and fluid. To keep one's place or progress up the corporate ladder seems to need a lot of hard thinking.

Location: Democratic Republic of Congo
Habitat: Deciduous rainforests, also grasslands and woodland
Population: c.20,000
Status: Endangered

Bonobo

Bonobos used to be known as pygmy chimpanzees and were once thought to be no more than a subspecies of their larger relative. Physically bonobos are very similar to chimpanzees, if a little less sturdy. Behaviourally, however, they could hardly be more different. The easy-going, peacefully erotic lifestyle of bonobos makes ordinary chimpanzees look like a bunch of rowdy hooligans.

The few thousand bonobos surviving today all live in the forests of what was once called Zaire, now the Democratic Republic of Congo, separated from the range of ordinary chimpanzees by the mighty River Congo. This natural barrier, across which neither species can swim, has prevented the two apes from meeting or mixing, even though, tantalizingly, they might be able to hear one another's cries across the water. This isolation allowed the bonobos' ancestors to evolve slightly differently from chimpanzees. Through a process called neoteny, in which juvenile characteristics survive into adulthood, they developed a slighter, more slender build. Today their shoulders are narrower, their necks thinner, their heads and ears smaller, their nostrils wider, and their eyebrow-ridges less prominent than those of chimpanzees. Also, the breasts of the females are slightly more rounded and the vocalizations generally higher-pitched.

Sexual relations

Bonobos also developed a more playful, more cooperative personality than chimpanzees. In addition to reproductive activity, they also began to use sexual actions in a playful, friendly way, as a means of strengthening social bonds within their groups. They also diverged in their competitive behaviour. A typical young male chimpanzee, as he progresses into adulthood, starts to bully younger males, compete noisily with older ones, and dominate all the females around him. Chimpanzee society is essentially patriarchal. Bonobo society, by contrast, is largely matriarchal. Female bonobos work together in groups to keep the males in their place. This difference in social organization is possibly related to the absence of gorillas south of the River Congo. North of the Congo, gorillas monopolize much of the edible vegetation at ground level, and the chimpanzees rely more on arboreal sources of food. To gather enough of

Take five
A bonobo male grabs a short break to sunbathe and rest before moving on to the day's next task, which could be feeding, grooming or, of course, a spot of sex.

Hmm
Compared to the often-aggressive chimpanzees, bonobos are relaxed and easy-going, preferring cooperation to conflict.

this kind of food, in a world dominated by the bigger, more aggressive, gang-forming males, the female chimpanzees must spread out. South of the Congo, in the absence of gorillas and chimpanzees, the bonobos have both arboreal and ground-level food sources available to them, so females do not need to split up to find enough to eat. This is an important factor in enabling them to form female alliances and empowering them to prevent male domination. One outcome of this is that the infanticide carried out by gangs of male chimpanzees is virtually unknown in female-dominated bonobo society.

It also means that, compared with chimpanzees, bonobos spend much more time on the ground, where sexual acts are more easily performed. Being at ground level for long periods allows bonobos to form large social groups, in which alpha-male domination would be virtually impossible to maintain. Communities of 50 to 200 bonobos have been seen living together in the wild, although most of the time these large groups are split up into foraging parties. Spending more time on the ground may also explain why bonobos have relatively longer legs than chimpanzees: they are in the process of becoming bipedal walkers, just like early humans. In fact, the latest studies of bonobos in the wild have revealed that they spend as much as 25 per cent of their time walking upright during periods of terrestrial locomotion. This alone makes them the most human of all our ape relatives.

Bonobos use sexual behaviour to settle disputes, diffusing the tension of the moment.

Make love not war

Another human-like characteristic of bonobos is that, unlike other great apes, the females are sexually attractive and active throughout most of their oestrus cycle. Bonobos use sexual behaviour to settle disputes. This has been put forward in the past as a unique feature of bonobo society, but the truth is that it has been known in baboons and some other primates for a long time. A female baboon, for example, will present her rump to a male as a peace offering; he may accept it by mounting, making a few perfunctory pelvic thrusts, and then dismounting. Similarly, a dominant male baboon may make a female sexual gesture to a scared, weaker male as a way of reassuring the subordinate. So there is nothing special about bonobo sexuality in this respect. What is unique about bonobos, is that, instead of simple, token copulation as a form of peace-making, they indulge in a wide variety of often very imaginative sexual contacts, underlining the fact that they are the most playful and inventive of all the non-human primates.

Some critics have expressed the opinion that the picture of bonobo society painted by current field workers reflects the present mood of Western society, and is a less than objective assessment. They point out that, ever since apes have been studied in the wild, the interpretations of field workers have been coloured by the cultural mood of the day. In the 1960s, for example, when Jane Goodall and Dian Fossey investigated wild chimpanzees and gorillas in detail for the first time, they described them in a way that fitted the mood of the 1960s flower children. Later, in the eighties, chimpanzees were portrayed as much more competitive and even warlike in their behaviour. Then, towards the end of the 20th century, when the bonobos were first studied in detail, their society matched rather well the post-feminist, sexually exploratory attitudes of the time. Although controversial, such criticisms may contain a grain of truth and should act as a warning to field workers to avoid being selective in their observations. That said, they cannot explain the very real differences that exist between the bonobo and the chimpanzee.

Baby face
A bonobo female illustrates the neotenous nature of the species compared to the chimpanzee – in other words, the retention of infant or juvenile features into adulthood.

Busy life
Compared to chimpanzees, gorillas, and orangutans, bonobo females are sexually active for a much longer part of their reproductive cycle – but everyone needs some time out.

Prosimians

For most people the word primate means monkey or ape, but this scientific order of mammals also includes some less brainy species, the prosimians, or 'before monkeys'. As well as having smaller brains relative to body size than the true monkeys and apes, the prosimians also have longer snouts and moist muzzles. Vision may dominate the world of the higher primates but for prosimians the sense of smell is still of great importance and they frequently employ scent in their territorial-marking activities, either depositing it from special glands or with their urine. The lower incisors of many prosimians form a comb that is used both for feeding and grooming the fur. Another distinguishing feature of this group is the long claw on the second toe of each hind foot.

ring-tailed lemur

Lemur

The ancestors of today's lemurs found their way to Madagascar, possibly on floating vegetation, just over 60 million years ago. In isolation, and free from heavy predation, these primates evolved into a fascinating and often beautiful group of animals. The most spectacular lemur, and the largest one alive today, is the ruffed lemur. It has black and white markings striking enough to rival those of the giant panda. Equally attractive is the ring-tailed lemur, with its black and white face-mask, its smooth grey body-fur and its long, bushy, conspicuous, zebra-striped tail. In addition to the large, typical, diurnal lemurs there are a number of shy, inconspicuous, nocturnal species with drab grey or brown fur. These are the much smaller mouse lemurs, dwarf lemurs and weasel lemurs, Madagascar's equivalents of bushbabies.

indri

Indri

The indri, the most extraordinary and largest of all the prosimians, is Madagascar's equivalent of an ape. Despite its conspicuous black and white markings, it is more often heard than seen. With the help of a special laryngeal air sac, its dog-like howls echo through the treetops, and one of its native names is 'dog of the forest' (the other, *babakoto*, means 'ancestor of man'). The indri is a spectacular leaper, sometimes covering as much as 9 metres (30 feet) in a single jump. On the rare occasions when it does come down to the ground, it moves along in awkward, bounding hops. For some reason, the indri does not survive in captivity, and it is never seen in zoos. It has some close relatives, the rare sifakas. These are even greater leapers, being aided in mid-air by a long tail that acts as a rudder. Sifakas spend most of their time in the branches, but when they do occasionally, come down to the ground, they move by jumping with their hind feet.

aye-aye

Aye-aye

The Madagascan aye-aye is one of the strangest and most specialized of all the primates. Looking like an oversized squirrel, but nocturnal in it habits, it feeds on fruits and the larvae of wood-boring beetles. With its enormous ears it has an acute sense of hearing and locates the burrows of beetle larvae by tapping on the wood with its long spindly fingers and listening carefully for any change in timbre. Once a burrow is found, it is gnawed open with the sharp, chisel-like incisor teeth and the grub is hooked from its hole with the skeletal, extra-long middle finger. The aye-aye is one of the world's rarest primates. Its total population is now thought to be little more than 2,000, with very few in captivity.

slow loris

Loris

The pop-eyed, Asiatic lorises look like forlorn, tailless bushbabies. The name 'loris' comes from the Dutch word for clown. Lorises advance along a branch with careful, deliberate steps in search of insects and other small prey. During the day they sleep curled up in hollow trees or on high branches. In the evening they wake up and go through an elaborate grooming session, before urinating on their hands and feet, with the result that, as they walk, they leave a scent trail behind them. Their equivalents in Africa are their close relatives the angwantibos and pottos. The potto has vertebrae with outward-pointing spines that protrude through the skin on the back of its neck. Although hidden by the animal's fur, they can be used in defence against a predator, the potto lowering its head and thrusting its body forward like a butting goat or ram.

galago

Galago

The bushbabies or galagos of Africa are similar to the lorises but instead of creeping along the branches they scamper rapidly about like squirrels and are capable of making huge leaps through the air. Aiding them in their acrobatics are their long, bushy tails that act like rudders and help to give them great balancing skills. They rarely come down to the ground but, when they do, they progress by hopping along like small kangaroos. Like the lorises, galagos are nocturnal and also, like them, they urinate on their hands and feet, making it easy for them to scent-mark their territory.

western tarsier

Tarsier

The nine, very similar-looking species of tarsier come from the Philippines, Indonesia and other parts of South-east Asia. They all have enormous eyes set close together on an otherwise tiny flat face, large, naked, bat-like ears and a long, naked tail with a small tuft of hairs at its tip. Tarsiers are so-called because the tarsal bones in their feet are greatly elongated. The long fingers and toes end in enlarged adhesive pads that enable the animal to cling to vertical surfaces, rather like a gecko. Tarsiers are great leapers and catch their insect prey by keeping very still and then springing at them and grabbing them with both hands. As with so many other prosimians, tarsiers mark the branches in their territories with their urine.

New World monkeys

There are today at least 200 different species of monkeys and possibly dozens more. Roughly half of these are to be found in the Old World, from Africa across Asia to the Far East, and half in the New World, in the tropical forests of South and Central America. They range in size from the formidable mandrill of Central Africa to the tiny marmosets and tamarins of the Amazonian rain forests. Old World and New World monkeys differ. Unlike Old World monkeys, some New World monkeys (all of which belong to the suborder Platyrrhini) have prehensile tails, a few are nocturnal and in some, certain digits end in claws rather than flat nails.

douroucouli

titi

white-faced saki

uakari

Douroucouli

Sometimes referred to as night monkeys or owl monkeys, the douroucoulis are the only nocturnal monkeys surviving today. In the New World, where there are no prosimians, they fill the niche occupied in the Old World by the bushbabies. Like bushbabies, they pass the day asleep in hollow trees, emerging at dusk to forage. They feed mainly on plant matter but also hunt for insects and small mammals. The males share the duties of carrying the young, and territories are scent-marked with urine.

Titi

Titis are very small, slow-moving monkeys that inhabit the rainforests of the Amazon basin. When resting on a branch they adopt a characteristic posture, with the body hunched, the tail hanging straight down, and the four feet placed close together. The fur is long, bushy and lustrous. Titis are very vocal monkeys, and their early morning concerts can be heard over huge distances, despite their small body size. As with some other small New World primates, the father carries the young when they are not being fed by their mother.

Saki

Sakis are untidy-looking animals. They have a cape of long hair over their shoulders and forehead, and a thick, bushy tail. Their strangest anatomical feature is their unusually wide nose. In some species the sexes are conspicuously different in appearance, the males having a vividly pale face, contrasting strongly with their black fur. Inhabitants of the tropical rainforests of the Amazon basin, sakis are fruit-eaters, but occasionally take small birds and mammals. In captivity they are gentle, sensitive animals, highly susceptible to shock. They rarely survive for very long away from their natural forest homes and are almost never seen in zoos.

Uakari

Mark Twain once said that humans are the only animals that blush, or need to. He clearly had not met the red uakari, whose bare-skinned head changes from a pale red to a deep scarlet when the animal becomes agitated. But, to be fair, this is more like a flush of anger than a blush of shame. Uakaris are small, short-tailed, chestnut-haired, crouching figures, looking not unlike rather bald and miniature orangutans. Even when young they give the impression of being elderly and depressed. In the wild they are restricted to the banks of rivers in the upper Amazon region, where they live only in the trees, feeding on fruits, buds, seeds, and leaves. They are hardly ever seen on the ground.

squirrel monkey

Capuchin

Although ungainly and lacking in the slender elegance of so many of their relatives, the prehensile-tailed capuchins make up for their rather ordinary appearance by being the most intelligent of all monkeys. They have been observed using tools in the wild, cracking open crabs and shellfish with stones, and in captivity have even been known to paint pictures. Because of their intelligence and their adaptability, capuchins have been kept as pets for hundreds of years. The name capuchin refers to the markings on the head that are reminiscent of the cowl of Franciscan (Capuchin) monks.

howler monkey

Squirrel monkey

These tiny, yellowish-coloured monkeys, with their attractive face-masks, are the most common primates in the New World. They live in large troops and are highly sociable. When sleeping, they huddle together, with their long tails wrapped around their bodies and their heads tucked between their knees. They feed on fruit and insects and prefer to scamper rather than leap through the trees.

Howler monkey

The prehensile-tailed howlers are the biggest of the South American monkeys, famous for their loud, booming calls made with the help of an enlarged larynx. Usually they join together to render a dawn chorus and perform another at dusk, announcing their territorial presence to rival groups. Their calls can be heard from a distance of some 3km (approximately 2 miles). Howler monkeys are generally rather placid, using howling and branch-shaking as a substitute for direct aggression against rivals. They are found in the upper levels of the dense rain forests of Central America and in the northern, tropical regions of South America, where their diet is mainly leaves.

spider monkey

Spider monkey

The gangling spider monkeys, with their long arms and legs, and their prehensile tail acting like a fifth limb, move through the trees like gigantic hairy spiders, hence their name. Their diet consists mainly of fruits, so they do not compete directly with the leaf-eating howlers that share their range. Spider monkeys rarely come to the ground but, when they do, are capable of bipedal walking with their tails curved up parallel with their backs. Unusually for a primate, the female is slightly larger than the male. She also has an elongated clitoris that looks like a penis.

woolly monkey

Woolly monkey

The compact, woolly fur of these monkeys gives them their name and their characteristic, stocky appearance. On the belly region they have longer, tufted, darker hair. They are closely related to the spider monkeys but their diet is broader, including both fruits and leaves. They are powerful, well-muscled monkeys but are not great leapers. When standing bipedally on the ground they have been seen to stiffen their muscular, prehensile tails and employ them as the third leg of a tripod.

Marmoset and tamarin

These, the tiniest of all the monkeys, have claws instead of flat nails on 18 of their 20 digits. The two exceptions are the big toes, each of which has a flat, monkey-like nail. There are more than 30 species, most in the forests of the Amazon basin. They feed on insects and plants, communicate with high-pitched, bird-like twitters and squeaks, and mark their territories with scent. The males share the duties of carrying the young. Marmosets and tamarins are very similar, but tamarins have slightly larger bodies, with smaller incisor teeth and longer canines. Marmosets use their longer incisors as 'tooth combs'.

Old World monkeys

Japanese macaque (snow monkey)

gelada

baboon

The Old World monkeys are part of the same scientific group of primates as the great apes, Catarrhini. This fact is telling: they are more closely related to the great apes, including our own species, than they are to New World monkeys. Old World monkeys differ from New World monkeys in several ways, the most obvious being their nostrils, which point downward rather than opening sideways.

Macaque

The Asiatic macaques are the least elegant of all the monkeys. Mostly stocky and with mousy, plain coats, they are widely distributed across southern Asia, from Pakistan to Indonesia. Opportunistic animals with few specializations, they are found in both highlands and lowlands, in dry habitats and humid ones, in the hot tropics and in cooler temperate regions. One of them, the rhesus macaque, is so common that it was seen as the obvious choice of subject for laboratory experiments. Macaque tails are sometimes short and may even be absent altogether. Two of the tailless species have been called apes in the past, namely the crested black ape of Sulawesi (formerly Celebes) and the Barbary ape of North Africa and Gibraltar. The Barbary ape is the only macaque found outside Asia, and the famous Barbary apes on the Rock of Gibraltar are the only wild-living monkeys in the whole of Europe.

Gelada

The gelada of Ethiopia is a heavily built, ground-living monkey, closely related to the baboons and with similar bright patches of skin. The snout is remarkably short, allowing the mobile upper lip to turn inside out in a dramatic expression called the lip-flip. Uniquely, the female gelada has a bare red patch of skin on her chest, surrounded by chains of bead-like swellings. These display-patches vary in size as she goes through her reproductive cycle, becoming redder and more swollen in unison with the reddening and swelling of the genital region. It is as if the chest of the female gelada mimics her genital region, allowing her to signal her sexual readiness from the front as well as from the rear. This mimicry is enhanced by the fact that her nipples have moved to the centre of her chest, making them look like copies of her labia and incidentally allowing her offspring to suck from both nipples at once.

Baboon

The African baboons are powerfully built monkeys with heavy heads, large elongated muzzles, and huge canine teeth. They spend most of their time on the ground, travelling in large groups. If threatened by a predator, the large male baboons cooperate with one another to form a formidable defensive gang. At night, however, they and the rest of the troop sleep up in trees for protection, or, in the case of the hamadryas baboon, on rock faces and cliffs. The hamadryas is sometimes called the sacred baboon because it was the species that was sacred to the ancient Egyptians. Over the years, thousands of mummified baboons have been discovered by archaeologists, buried in special baboon cemeteries.

Mandrill

The West African mandrill, and its close relative the drill, used to be thought of as unusual baboons, but it is now recognized that they are more distinct, and they have since been given their own separate genus in primate classification. The brightly coloured mandrill is the largest of all monkeys. It lives among the rocky outcrops that occur in the depths of Africa's rainforests, and it spends a great deal of its day on the ground. Like the gelada, the mandrill displays genital mimicry, but in this case it is the male that has a copy of its genital region. Its bright red nose and blue cheeks are mimics of its red penis and blue scrotal patches. The slightly smaller drill lacks these head colours, having a black face fringed by a white ruff.

mandrill

Guenon

These are the slender, long-limbed, long-tailed African monkeys, such as the vervet, the grivet, the mona, the white-nosed and the green monkey. They are found all over sub-Saharan Africa and include the most common and widespread of African primates, although a few of the species are now rare. Most guenons have vivid facial markings that act as species 'flags'. The majority live at the edges of forests, but some, like the patas monkey, are almost entirely ground-dwelling and live in more open habitats. The patas can run on the ground faster than any other monkey. The smallest of all the guenons and, indeed, one of the smallest of all the Old World monkeys is the talapoin. Sometimes called the pygmy guenon, it lives in the swampy forests of West Africa.

Mangabey

Like the guenons, the mangabeys are long-limbed, slender-bodied monkeys found in tropical Africa, but they are larger and are confined to the west of the continent. They do not compete with the guenons, partly because their large size and powerful jaws allow them to feed on fruits with exteriors too tough for guenons to penetrate. Mangabeys live in troops, which are led by one or more large adult males.

black and white colobus monkey

Colobus monkey

Colobuses feed entirely on leaves. They are the most arboreal (tree-dwelling) of all African monkeys. Their name is derived from the Greek word for mutilated and refers to the fact that, unlike other monkeys, they almost or completely lack thumbs. There are several species of colobus monkeys but by far the best known and most spectacular are the two species of black and white colobus, with their boldly contrasting coats of black and white hair.

Langur

The leaf-eating monkeys of Asia are known collectively as the langurs. Found all the way from Nepal and India through to Sumatra, Java and Borneo, they include in their number the extraordinary, bulbous-nosed proboscis monkey. Typical langurs are slender, long-limbed monkeys with opposable thumbs that supplement their largely leafy diet with a little fruit and some flower buds.

proboscis monkey

Gibbons

white-handed gibbon

The gibbons are highly specialized primates which, because of their smaller size, are often set to one side when apes are being considered. In the popular imagination they are thought of almost as tailless monkeys rather than as diminutive apes, but this is a mistake. There is evidence that their recent ancestors were much larger, and there can be no doubt that, if a giant gibbon existed today, it would receive the same amount of attention as the other apes. Also, the gestation period of a gibbon is more like that of another ape than a monkey, suggesting that their reduced size is comparatively recent in evolutionary terms and no more than an adaptation to rapid movement through the treetops. Bearing this in mind, it is worth taking a closer look at these remarkable Asian acrobats.

There are several different species of gibbons to be found hanging around in the treetops of South-east Asia, but they are very similar to one another in the way they organize their lives. They are all long, lean, lanky primates with amazing leaping and swinging abilities, capable of sailing through the air from one branch or tree to another with a casual elegance that makes them a joy to watch.

Gibbons live in smaller family groups than the bigger apes, and although there is some local variation, the basic arrangement is for a monogamous pair to keep together and not to join up with other adults. Gibbons are territorial, announcing their presence not by visual displays but by ear-piercing cries. Their extraordinary, whooping songs, which can be heard over huge distances, even in the thickest forests, are unique and unforgettable. One species of gibbon, the siamang (see opposite), has a large inflatable throat pouch that amplifies these sounds even more.

Singing swingers

With their pair-bonds, their songs, and their flitting from tree to tree, the gibbons are the most bird-like of all the primates. Inevitably, they have also proved to be the most difficult for close observation in the wild. What they lack in visibility they make up for in audibility. Every day the male and female of a pair will sing their song as a way of affirming their bond of attachment and at the same time as a way of advertising their ownership of a particular territory. These songs are unusual, in that they are duets with the male and female taking different roles.

Gibbons are also unusual in that they do not have a midday siesta. Other primates typically feed in the morning and the evening, with a long rest in the midday heat. Gibbons remain active throughout the middle of the day, but then retire early, several hours before sunset. Unlike the great apes, gibbons do not make sleeping beds in the trees. They may lie down on a broad branch, but more usually they sleep in a curled-up sitting posture with their knees bent up to their chins, their hands folded on their knees and their faces buried between their knees and their chest. This tightly hunched posture helps to keep them warm, and their thick fur protects them from downpours of rain. Like monkeys that sleep in a similar posture, gibbons possess hard, thickened, hairless skin pads called ischial callosities on their rumps. This is something that the bed-making great apes have lost.

Particular tastes

When it comes to their diet, gibbons are specialists, focusing most of their attention on fruit pulp. Their favourite fruits are figs and other small, colourful, sugary, and pulpy fruits. In their search for suitable food sources, gibbons travel about 1km (0.6 miles) a day, although some have been known to cover over 3km (1.8 miles). The home range of a gibbon family is slightly larger than its rigidly defended territory. On average, about three-quarters of the home range is kept as an exclusive area for the resident family. In other words, there is a defended core to the home range and outside that the family may venture from time to time in search of extra food sources. The adult pair usually has one or two offspring with them at any one time, and the very young ones are particularly adept at clinging tightly to their mother's fur.

Authors disagree about exactly how many kinds of gibbons there are, with some recognizing only five main forms and others claiming that these five should be subdivided into twelve or more species. Among the main forms are the following:

Siamang (*Symphalangus syndactylus*)

The siamang, found in Malaysia and Sumatra, is the largest of the gibbons, and has a long, shaggy coat. The fur is black except for a pale patch around the mouth and chin. The arms are even longer than in other gibbons, the body is heavier, and the hands and feet are broader. The second and third toes of each foot have webbing between them. There is an inflatable throat sac used as a resonator that creates a booming sound during bouts of loud singing. The song of the siamang is the one of the loudest sounds made by any animal on Earth.

Hoolock gibbon (*Hoolock hoolock*)

This gibbon, found in eastern India, Bangladesh, Burma, and southern China, is distinguished by its white eyebrows, which, in the male, show up vividly against its brownish-black fur.

Concolor, crested or black gibbon (*Nomascus concolor*)

This gibbon, found in southern China, Vietnam, Laos, Cambodia, and Thailand, has an erect crest of hairs on the head that distinguish it from all other species. In concolors there is a colour difference between the males, females, and young: adult males are black, females buff, and the young a pale yellowish-grey. There are also two local colour forms in which the males' cheeks and throat are either buff-white (*N. c. leucogenys*) or reddish-buff (*N. c. gabriellae*).

Dwarf or mentawi gibbon (*Hylobates klossii*)

This gibbon, also called Kloss's gibbon, is found only on the small Mentawi (Mentawai) Islands, just to the south of Sumatra. It was originally called the dwarf siamang because it too has a vocal sac, but it is now believed to be more closely related to the lar gibbon.

White-handed or lar gibbon (*Hylobates lar*)

The white-handed gibbon is found in Malaysia, Thailand, China, Borneo, Java, and Sumatra. In the typical form it has a ring of white fur around its dark face, and pale hands and feet, but there are several colour variants, such as the black-handed gibbon (*H. l. agilis*), the pileated gibbon (*H. l. pileatus*), and the grey or Javan gibbon (*H. l. moloch*).

siamang

hoolock gibbon

Of Apes and Men

Our relationship with our wild cousins has changed greatly over the last 50 years. Most of us no longer consider all animals 'fair game', to be shot for sport or captured and used for our entertainment. Thanks to the pioneering work of field scientists such as George Schaller, Jane Goodall and Dian Fossey, we have gradually come to learn more about the great apes, acknowledge our close relationship with them and treat them with more respect.

Shared traits of the great apes

Every human culture on Earth has some kind of explanation as to how it arrived here. One tribe claims that the Great Earth-mother gave birth to humans and all other forms of animal and plant life. Another believes the first men rose up from the ground. Yet another imagines the first people were created by an all-powerful supernatural being or 'god'. There are endless variations, all concocted to explain something that was beyond their understanding. Some of these myths are still with us, but in the Western world the concept of evolution, introduced by Charles Darwin in the 19th century, has replaced the many legends of yesterday.

Set free by Darwin's pioneering work, the biological and medical sciences have made such enormous strides in the 150 years since the publication of Darwin's most famous work, *On the Origin of Species*, that we now have a very clear picture of the animal nature of the human species. It is obvious enough that the skeletal details of the human body are very similar to those of the great apes. True, some bones are heavier or longer, while others are lighter or shorter, but the overall design is the same and this leaves little doubt about our close affinity. The same is true of the internal organs and the sense organs that we and the great apes possess. We also have the same kind of immune system, nervous system and blood. Even our fingerprints are similar.

Alike in actions

Humans and great apes, especially chimpanzees, also show many striking similarities in behaviour. They are both capable of rapid one-trial learning and both experiment with various kinds of tool-making and tool use. They both display the ability to pass local traditions on from one generation to the next. If a particularly bright ape invents a new way of obtaining food, others will quickly imitate it, leading to different regional trends. Both apes and humans employ subtle strategies to gain status and control power in the community. When two rival groups come into contact they are both liable to engage in confrontation or attack. Chimpanzees have even been observed to engage in what might be called primitive warfare. Both humans and chimpanzees will hunt and kill other species to obtain meat and, when successful, will respond to a sudden food surplus with food sharing. There are even 'feminist' female apes that have been known to set up alliances with other females in order to achieve dominance over a stronger and more powerful male.

Genetic studies have confirmed just how close we are to the other apes, and taxonomists have been forced to place orangutans, gorillas, chimpanzees and the human species all in the same family, the Hominidae. The old arrangement saw the hairy apes in one family, called the Pongidae, and our own species in another called the Hominidae. Now the Pongid family has been abandoned.

Helping hands
(Previous page)
Less hair on the hands than on the arm, fingernails, fingerprints, and fingertips, and a delicate sensitive touch. Orangutans' hands are so similar to ours – note the orangutan's smaller thumb set high towards the wrist.

Common behaviour
The common chimpanzee wields a stick in a pose that demonstrates it shares with humans the advantages of bipedalism.

Playtime
Just as we toss our children around when indulging in physical games, this bonobo mother is having inventive fun with her offspring.

The main reason for this change has been the information obtained from detailed studies of chimpanzee and human genomes. A report in 2005 stated that: 'the first comprehensive comparison of the genetic blueprints of humans and chimpanzees shows our closest living relatives share perfect identity with 96 per cent of our DNA sequence.' Earlier reports had put the figure as high as 98.8 or even 99.4 per cent. Whichever version we accept, with this new genetic knowledge it is hard to see how the human species can ever again be viewed as some sort of mystical 'special creation', completely separate from the rest of nature.

Closer cousins than once thought

Perhaps even more significantly, it has also been claimed that humans are more closely related to chimpanzees than chimpanzees are to gorillas. If the evolutionary gap between humans and chimpanzees is indeed as small as this, it becomes necessary to reconsider the classification of the chimpanzee. It has been suggested that the chimpanzee should be placed in the same genus as humans – the genus *Homo*. In New Zealand a few years ago a group of zoologists and lawyers petitioned the country to pass a bill conferring rights on chimpanzees that would see them treated legally in the same way as humans. This would mean that experimenting on a chimpanzee would be viewed as torture, eating chimpanzee flesh as bushmeat would become cannibalism, and killing a chimpanzee would be murder. Critics of this view asked whether, if a chimpanzee bit a man, it would be tried and sent to jail for assault. This brought the concept into sharper focus and revealed its weakness, but the New Zealand government did at least grant great apes legal protection from animal experimentation.

If the proposed Bill of Ape Rights were to be fully accepted worldwide it would create a huge problem for zoologists. All captive apes, imprisoned without trial and innocent of any crime, would have to be released immediately and returned to their wild habitats. This would include all zoo apes and would destroy the carefully designed breeding colonies that major zoos are attempting to establish all over the world. However, as it is formulated at present, the proposed Bill of Ape Rights overlooks three important issues. First, it ignores the fact that the wild environments of the great apes are under severe threat from logging, mining and other human activities, and may eventually disappear altogether. Those environments have also become much more dangerous places for apes to live, due to increases in poaching and hunting for bushmeat. Second, it begs the question of where the thousands of captive apes would be let loose. Even if a vacant patch of forest could be found for them, the vast majority would find it hard to readapt to the wild state after years in captivity. Third, if all the great apes at present living in zoos and safari parks were returned to the wild, their species might soon become 'out of sight and out of mind'. The presence of a few great apes in zoos and animal parks, providing they are kept in large enclosures in natural social groupings, can act as a valuable, living reminder of how exciting these species are. They are ambassadors for their species, keeping public interest alive and reinforcing the need for urgent conservation.

The presence of a few great apes in zoos and animal parks can act as a valuable, living reminder of how exciting these species are.

Home from home

Instead of banishing all zoo occupants to an uncertain future in the wild, a registered list of all captive apes could easily be made. Those licensed apes could then be provided with appropriate captive conditions and continue to breed and educate the general public. It would, however, be required that all captive apes be given conditions that mimic their wild habitats as closely as possible – it would become a crime to keep a chimpanzee in solitary confinement in a small, bare cage, for example. Now that we know how very closely we are related to them, such rights are the least they should be given. If this feels like a compromise, it should be pointed out that, in a hundred years time, the descendants of those captive apes might be the only ones left alive on the planet, with their natural habitats having been flattened to make way for the ever-growing human population. A future in captivity may not be ideal, but it is better than no future at all.

That's us!
Mirror self-recognition is known in chimpanzees and bonobos as well as in orangutans; it has also been observed in elephants and dolphins, and in some birds such as magpies and kea parrots.

The story of DNA

DNA is nature's instruction manual. It contains the codes for building all the parts that make up living things. DNA occurs in nearly every single cell in the human body and, for that matter, in almost all cells in nearly all living things on Earth. In the cells of animals such as ourselves most of the DNA is contained inside a structure called the nucleus, which forms a ball-like core to every cell. Cells are the microscopic structural building blocks of animals. Some, such as blood cells, are mobile and exist apart from others, but most have a fixed position in the body and are surrounded on all sides by others like themselves, rather like individual stones or bricks in a thick wall.

The existence of DNA, or something like it, was long suspected by scientists. As far back as the 19th century the study of genetics had already begun, albeit on a small scale by individual pioneers such as the Austrian priest and scientist Gregor Mendel (1822–1884). Mendel went on to show how certain physical traits were inherited and passed down through the generations. Mendel's work was largely ignored until about 1900, but then scientists began to investigate inheritance further and talk about genes – sections of biological code that carried the instructions for the manufacture of particular physical characteristics in the body, such as eye or skin colour, among other things.

What we now know is that it is DNA that carries those instructions. Genes are essentially small sections, or chunks, of DNA, which is itself an extremely long chemical molecule. The letters DNA stand for deoxyribonucleic acid. This complicated sounding name belies the essentially simple structure of the molecule itself. Put simply, DNA is a chain formed from two long, intertwined, corkscrew-like 'backbones', the double helix, carrying varying combinations of just four chemical units known as bases. It is the order in which these units appear in the chain that forms the so-called genetic code, the code that instructs the construction of the chemicals that form the cells and bodies of living things.

Family resemblance

The complete set or genome of DNA for every species of living thing is different. Even within species there are very subtle differences between the DNA of different individuals. The degree of difference between the DNA of individuals is a reflection of how closely or distantly related they are to one another. Siblings have more similar DNA than cousins and cousins have more similar DNA than individuals who are not related to one another at all. By the same token, the degree of difference between the DNA of different species also shows their relatedness to one another. Species with similar DNA are usually more closely related than species with DNA that is very different.

Ever since the human DNA sequence was determined, projects have been undertaken to establish the DNA sequence of other animals.

Our closest relatives are other humans that are now extinct. All that we have to identify them are fossils. In most cases their relatedness to us has been inferred by physical similarities, particularly the size and arrangement of the teeth and the shape of the skull. Recently, however, small fragments of DNA have been successfully extracted from the remains of Neandethals, our very close relatives that became extinct perhaps no more than 27,000 years ago. Incredibly that DNA has been sequenced and its code compared to that of our own species. Perhaps unsurprisingly, the Neanderthal DNA code is very similar to that of modern humans, but it does contain a few significant differences. Overall, say the scientists who worked on the project,

Early development
Researchers are identifying genes that cause differing speeds of development between humans and chimpanzees, for example, in brain growth during the first months.

Neanderthals were three times as different from us as we all are from each other. Obtaining DNA from living great apes is much simpler than extracting it from extinct ones – a blood sample or even cheek swab is more than sufficient. Ever since the human DNA sequence was determined in 2003, projects have been undertaken to establish the DNA sequences of other animals, including the wild great apes. The most comprehensive of these is the Chimpanzee Genome Project. Several important discoveries have already been made. As one might expect, the genes that encode speech development in humans are very different in chimpanzees. Several genes involved in hearing are also rather different. This reflects the differences in human hearing from that of the other great apes. All in all, the sections of the chimpanzee DNA sequence examined so far suggest that chimps are around 10 times as different from us as we are from each other. In other words, for every one area on the human DNA sequence that differs between individual people there are 10 areas that are different on the DNA sequence of a chimpanzee. While this may sound a lot, there are many more areas where the DNA sequences of chimpanzees and humans are identical. Early results from the Chimpanzee Genome Project suggest that some 96 per cent of the chimpanzee DNA sequence is a perfect match with our own. More recent studies suggest that that figure is even higher.

Extinct ancestors

Our own species is by far the most successful and widespread of all the great apes. Today there are more than 6,600 million of us spread across the planet, living almost everywhere, from the tropics to the poles. Our beginnings, however, were much more humble. Just 10,000 years ago, when we had already spread from Africa through Europe and Asia, and pioneering souls had reached the Americas and Australia, it is thought that the entire human population numbered no more than 5 million individuals. Before that, our numbers were much smaller.

Our species, *Homo sapiens*, is the only survivor of the human lineage. We have been alone on the planet for 12,000 years, since the last remaining members of the dwarf species *Homo floresiensis* (nicknamed 'the hobbits') died out in Indonesia. Fossil evidence, however, shows that we and the 'hobbits' are far from the only humans to have inhabited the Earth (human is here taken to mean great apes that were more like our own species than they were like any of our wild, living cousins). In fact, to date there are up to 22 extinct human species known. At some periods there were several living alongside one another. The human lineage is rich and dates back some 7 million years.

It began in Africa

The earliest known member of the human group, known as the hominins – the trunk of our family tree – is a species that goes by the name of *Sahelanthropus tchadensis*. This creature lived in what is now the country of Chad, which borders Sudan in the northern half of Africa. It is known from a single skull – the Toumai skull, two fragments of jawbone, and three solitary teeth. These fossils came from five separate individuals and have been dated at seven million years old. Other fossils found alongside them are mostly of extinct animals, including giant saber-tooth cats, four-tusked elephants, and three-toed horses.

Fossil evidence suggests that we and the 'hobbits' are far from the only humans to have inhabited the Earth.

Sahelanthropus differs from the hairy apes and their ancestors in its dentition, thick, continuous brow ridge, and the point at which the bones of its spine joined its skull (much farther down, towards the base). This latter feature suggests that it may have habitually walked on two feet, although until more skeletal remains are discovered this point remains speculative and debatable. The continuous brow ridge suggests that it was a direct ancestor of our own species, since this feature is known only from our own skulls and those of other members of the scientific genus *Homo* (such as the Neanderthals, *Homo neanderthalensis*), to which we belong.

The comparison of features of *Sahelanthropus* with modern humans makes an important point: not all of the extinct human species were direct ancestors of ourselves. Many, although close relatives, formed separate branches of the human family tree, branches that eventually petered out into nothing. Among them were creatures such as *Paranthropus aethiopicus*, a sort of human gorilla. It lived in equatorial eastern Africa from 2.8–2.3 million years ago and like true gorillas fed almost entirely on vegetable matter.

Following in Lucy's footsteps

Our own genus, *Homo*, appeared perhaps more than two million years ago. Before that time (but after the extinction of *Sahelanthropus*) the majority of known human species belonged to the genus *Australopithecus*. The best-known and most studied species of *Australopithecus* is

Australopithecus afarensis, which lived in what are now Ethiopia, Kenya, and Tanzania, from about four to just less than three million years ago. It is this species to which the famous Lucy belonged, a 40 per cent complete skeleton discovered in 1974 by the American scientists Donald Johanson and Tom Gray. *Australopithecus afarensis* probably walked upright, and it is more than likely that members of this species made the Laetoli footprints. The various *Australopithecus* species were human in as much as they were more closely related to us than to the other great apes, but whether or not they were our direct ancestors remains uncertain. Closer to us in relatedness and time are the other members of our own genus, *Homo*. One of the earliest of these, *Homo rudolfensis*, is known only from fossil skulls and jawbones. These date from about 1.9 million years ago and, like most of the earlier human fossils, come from eastern Africa, specifically Kenya. Interestingly, not long after this time other species of *Homo* began appearing in Asia. Among them were *Homo georgicus*, named after the country of Georgia where it was discovered, and *Homo erectus*, first identified in Java in 1891 as 'Java Man' but since known from many fossils across Africa and Asia, and in southern Europe too. The distance separating these species from their African cousins suggests that humans were by this time much more widespread than the dots on the map representing their few fossil finds might otherwise lead one to believe.

Our species, *Homo sapiens*, appeared less than a quarter of a million years ago – exact dates are still fiercely debated in the scientific community. Although a few argue otherwise, our species almost certainly arose in Africa and spread outward from there. We were not the first humans in Europe – members of the species *Homo heidelbergensis* (which probably later gave rise to *Homo neanderthalensis*) had made it as far as Britain some 500,000 years ago – long before *Homo sapiens* had even appeared. However, we were the first and indeed only humans to colonize the Americas and Australia. For that matter, we were the only apes on those continents.

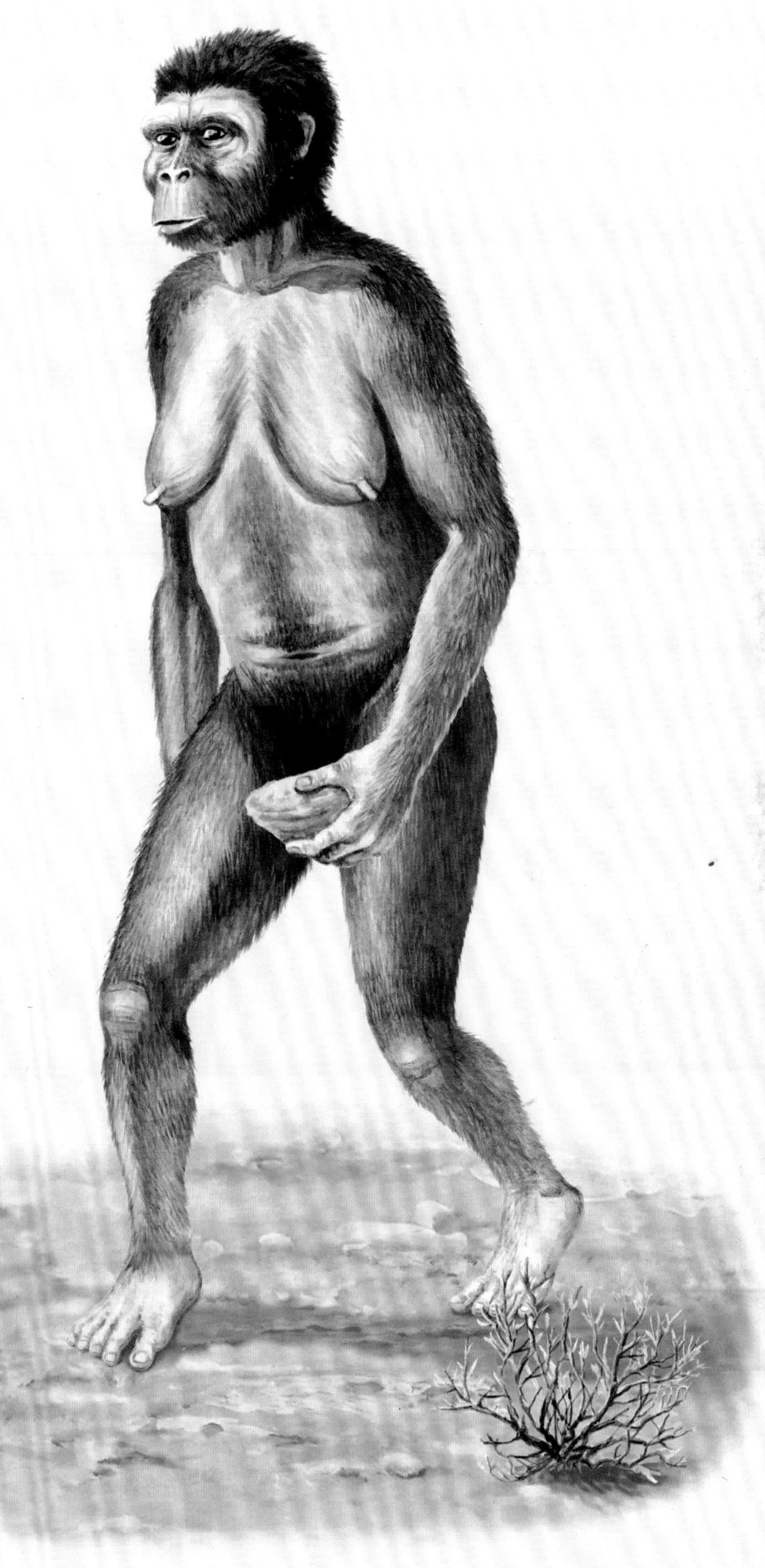

Lucy
The partial skull, spine, pelvis (hip bone), and leg bones of 'Lucy' (*Australopithecus afarensis*) suggest an almost upright posture and bipedal walking. 'Lucy' fossils are dated to just over three million years old. With a height of about 1.1 metres (about 43 inches) this specimen indicates that upright walking appeared before an increase in brain size. Lucy's skull remains are sparse but suggest a brain size of less than 500cc and perhaps less than 450cc which is roughly comparable to a chimpanzee.

Living relatives?

↖ It's abominable!
The name yeti is derived from the Tibetan for 'rock bear'. In Westernized popular legend it appears mostly as some type of giant ape-like creature.

There have always been tales about great hairy beings lurking in the hidden corners of the world. There have even been claims that some of these monsters have been photographed or their presence verified by footprints. They have divided the world into believers and sceptics but, to the open-minded scientist, they remain a fascinating possibility.

Nothing could be more exciting for those who study primates than to encounter a large, unknown species that has hitherto avoided detection. Whenever a claim is made that such a being exists, primatologists look at the evidence. Sadly, the hard evidence produced so far is severely limited, and the beasts themselves are so elusive that no definite conclusions can be drawn. Those individuals who believe in the existence of these mystery apes pour scorn on the scientists who demand more evidence, saying that the experts are frightened of finding something that may upset their neat evolutionary family trees. Nothing could be further from the truth. Give any primate expert a piece of solid evidence – a bone, a skull, a piece of skin – and if it belongs to a hitherto unknown species of ape, they will go to the ends of the Earth to track it down. On the other hand, those who argue that all suggestions of the existence of mystery apes are romantic imaginings or cynical hoaxes, are displaying closed minds. A balanced approach, halfway between these two extremes, enables every new claim to be examined and classified as either Clever Hoax, Wild Fantasy, Ancient Legend, Mistaken Identity, or Possible New Species. With this in mind it is worth taking a brief look at three of the most famous 'mystery apes'.

The yeti of the Himalayas

The first reported sighting of a yeti to reach the Western world was recorded by the British Resident at the court of Nepal in the early 19th century. It was described as a frightening, hairy, tailless wild man. Later that century the first sighting of yeti footprints was made, in snow at 5,180 metres (17,000 feet) in Sikkim. Similar footprints were seen in 1921 at 6,100 metres (20,000 feet), near Everest, and in 1925 the creature itself was sighted on the Zemu Glacier at 4,570 metres (15,000 feet). In 1930 a journalist enlivened the story by inventing the name the Abominable Snowman, increasing the yeti's popular appeal. The most detailed early yeti sighting took place in 1942 on the border between Bhutan and Sikkim. Seven prisoners-of-war had escaped from their detention camp in Siberia and were trekking across the Himalayas to India. They spotted two figures walking through the snow and described them as being about eight feet tall, covered in reddish-brown hair, with squarish heads, very small ears, sloping shoulders, powerful chests, and long arms with wrists reaching to the knees. In the 1950s a number of Western mountaineers reported seeing yeti footprints. Among these was the celebrated conqueror of Mount Everest, Sir Edmund Hillary. Another famous mountaineer, Lord Hunt, not only saw the footprints of a yeti, but also 'heard [its] yelping cries' in the distance. With respectable Westerners of this kind making claims, it became increasingly difficult to ignore the possibility that the yeti existed, and whole expeditions were organized just to track it down. Needless to say, they failed.

Nothing could be more exciting for those who study primates than to encounter a large, unknown species that has hitherto avoided detection.

Hard evidence?
↗ An alleged yeti footprint near Mount Everest, 1951.
↗↗ Everest conqueror Sir Edmund Hillary shows a likeness from apparent eyewitness accounts prior to his yeti-hunting expedition of 1960–61.
→ Footprints attributed to the yeti, 1961.
→→ Photos and video sequences of Bigfoot or Sasquatch, supposedly the yeti's North American counterpart, are given little credence by scientists.

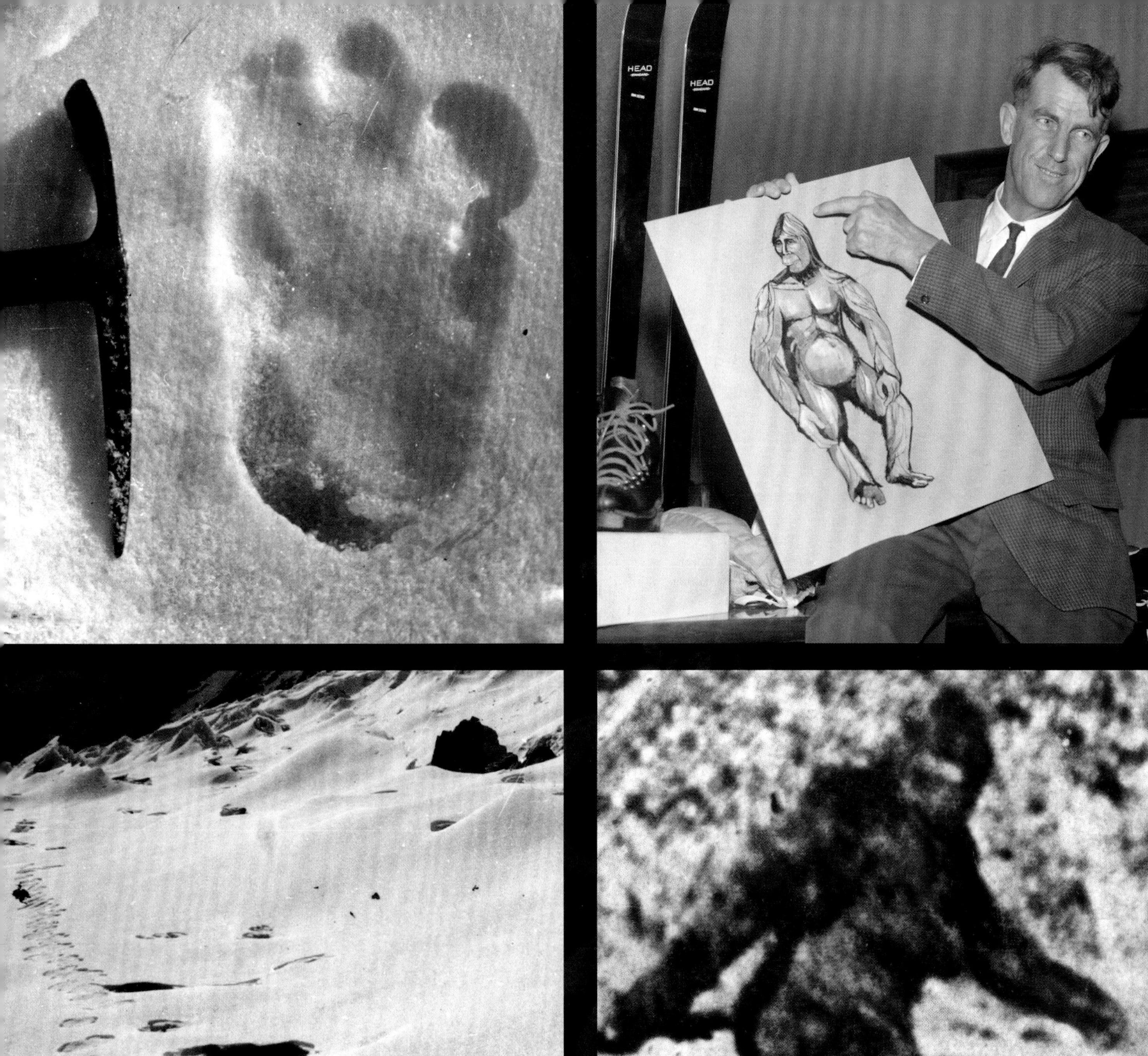
HEAD
HEAD

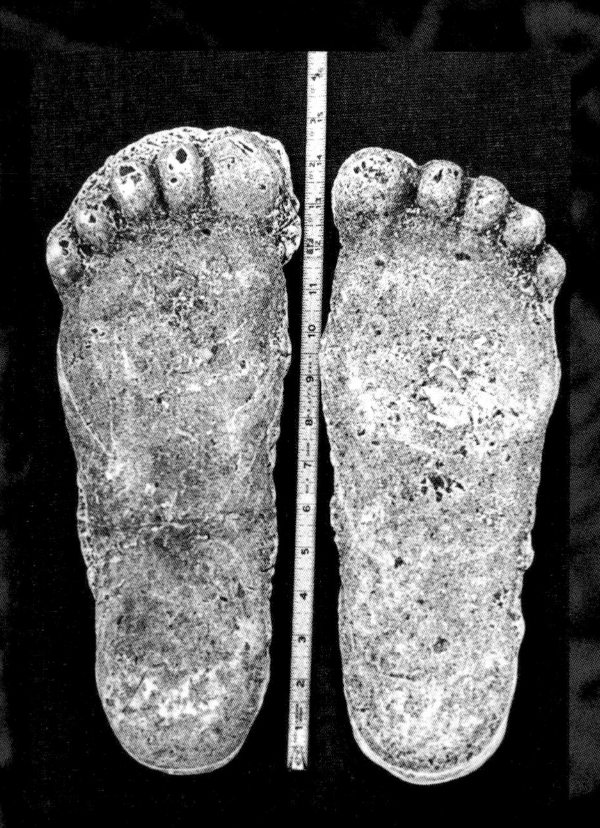

Bigfoot by name... Casts of 36cm (14½in) footprints found at Bluff Creek, California – thought by some to belong to Bigfoot himself!

Possible explanations

As interest in the yeti grew so did speculation over exactly what it might be. Some people thought that it might be a kind of bear. Others pointed out that some Hindu holy men were known to travel almost naked over snow to the caves where they lived out their ascetic lives. These suggestions were rejected by those who had personally observed yeti footprints. To them, there was no doubt that the yeti was real. Some suggested that the creature inhabited the densely forested valleys below the snowline and that the earlier sightings, higher up in the extreme cold, were of isolated wanderers from the main yeti communities. Many local people reported seeing yetis in these dense, subtropical valleys. There were even stories of yeti lairs and sleeping nests similar to those made by gorillas.

Reported sightings are one thing but hard evidence is another. At one point, however, it did look as if such evidence might exist. Some Buddhist monasteries near Mount Everest claimed to keep yeti scalps for use in sacred dances. Sir Edmund Hillary was allowed to bring one of these sacred artefacts to London to show to zoologists. Unfortunately the scalp turned out to have been made from the skin of a rare Himalayan goat-antelope called the serow. To date, that is where the yeti story ends. All of the evidence so far is circumstantial and could be dismissed as little more than travellers' tales and local legends. Only a few bizarre footprints remain to puzzle the sceptics and keep the story alive. Recent expeditions have produced more footprints and claims of sightings, but nothing has been added in the way of hard biological evidence. Were it not for the hundreds of supposed sightings by both locals and visitors, the yeti story would have been relegated to the world of imaginative fiction. Instead it must remain an intriguing footnote in the history of primate studies.

The bigfoot

If Asia has its yeti, North America can claim an equally improbable beast, the bigfoot. Both are the subject of hundreds of sightings and in both cases we have the evidence of huge footprints. Also, in both cases, there is much folklore attached to the mystery ape.

The bigfoot, or sasquatch, is supposed to be living quietly, minding its own business, in the heavily forested and uninhabited regions of the Pacific North-West. It was first reported as long ago as 1784, when *The Times* of London mentioned a huge, manlike, hair-covered creature that had been captured by local Indians. In late Victorian times, a similar beast was said to have been captured in British Columbia. It was described as being rather like a gorilla, covered in glossy black hair, much stronger than a man, and making barking or growling noises. It was shipped east, but died in transit, and there is no record of what became of the corpse.

Later, among hundreds of reported sightings, one was very precise about the size of the animal. In 1940, a Native American family living in a remote spot in British Columbia said that they had been visited by an 2.4 metres (8 feet) tall sasquatch that left footprints 40 centimetres (16 inches) long. Between then and the present day there have been countless similar reports, but no hard evidence. Indeed, the biggest weakness of the case of the bigfoot, as with the yeti, is that, given the huge number of claimed sightings, it seems inconceivable that no scientific material that could settle the matter once and for all has ever been forthcoming. Not a single corpse or even a bone of the creature has ever been found, for example. It is for this reason that most scientific authorities simply relegate the bigfoot and all the mystery apes to the realm of amusing fictions.

Caught on camera

Confounding the situation of the bigfoot is the very real possibility of hoax. In 1967 a film of a supposed bigfoot walking through a northern Californian forest was shot by a man called Roger Patterson. The figure in the film is covered in short, shiny black hair, has a bony crest on its head, a short neck, pendulous breasts, and heavy shoulder and back muscles. It was calculated to be about 2.1metres (7 feet) tall and left footprints that were 37 centimetres (14.5 inches) long. This was either the first hard evidence of the existence of a modern mystery ape or a clever hoax, probably perpetrated to attract tourists to the area, or simply to gain notoriety for the man who filmed it. To this observer, at least, there is something unnatural about the gait of the creature in the film that suggests an actor in a gorilla suit. And the photograph of Patterson holding up the plaster cast of a gigantic humanoid footprint is also hard to swallow. One can only hope that something more convincing appears in the near future.

Sumatra's mystery ape

For over a century there have been stories and rumours circulating in Sumatra about a mysterious relative of the Sumatran orangutan called the orang pendek. The name orangutan means 'person of the forest', or, if you prefer, 'the old man of the woods;' orang pendek means 'short person'.

No hard evidence for the orang pendek has ever been found, but the local belief in its existence is persistent. Native forest tribes are convinced that it is there, as are the local villagers. Sightings were reported by early Dutch colonists and, in more modern times, have been claimed by naturalists, travellers and explorers. Most of the recent sightings have taken place in central-western Sumatra, some distance from the northern territory occupied today by the surviving orangutans.

The orang pendek is said to dwell in the most remote, mountainous forests of Sumatra, where it has become ground-living and bipedal. It is thought to be about 1.5 metres (5 feet) tall, with hair that may be brown or dark grey. Its head is said to slope back to a distinct crest, and its face has been described as having a bony brow-ridge, wide-set eyes, a small mouth with broad incisors and long canines, and a humanoid nose. It is also said to have a huge chest; long powerful arms and short legs; and feet with a thumb-like, opposable big toe.

A case of mistaken identity

Apart from its humanoid nose and its darker coat colour, the descriptions of the orang pendek could easily fit any large male orangutan that had briefly come down to the ground. The opposable big toes suggest an arboreal adaptation, rather than a primarily ground-living existence. The problem with dismissing the orang pendek as no more than an unusual, local variant of the orangutan is that those who claim to have encountered it say it is strikingly different from the well-known red ape. Some have suggested that it might be a living relative of *Ouranopithecus*, an extinct primate that had characteristics of both modern humans and apes. Encouraged by reports of sightings, a 15-year investigation was begun in the early 1990s. Sadly, when the study ended, the existence of the orang pendek remained unproven. Since then, a plaster cast of a supposed orang pendek footprint has been examined by a Cambridge primatologist, who came to the conclusion that it was 'definitely [from] an ape with a unique blend of features from gibbon, orangutan, chimpanzee, and human'. Also, orang pendek hairs have recently been recovered, and detailed analysis has revealed that they come from 'a previously undocumented species of primate.' DNA tests, however, suggested that the hairs were human, or had been contaminated by human contact. Despite the DNA findings, these results were sufficiently encouraging to prompt a major new field study and this was begun in 2005. Even if the orang pendek turns out to be no more than an isolated subspecies of orang-utan, it will still be of enormous interest, should its existence finally be proven beyond doubt.

Uniquely human

How humans differ from the great apes

Human beings differ from the other apes in five important ways. They are naked apes, having lost their coat of fur. They are walking apes, having adopted bipedal walking. They are talking apes, having developed grunts into words. They are playful apes, having extended infant playfulness into adulthood. And they are brainy apes, having evolved the best brain on the planet.

The very first thing one would notice if a human body was laid out on a slab alongside all the other apes and monkeys is that it is the only one that has naked skin over most of its surface. True, some species have a naked chest, a naked face or a naked rump, but only humans have naked skin everywhere except for tufts above the genitals, in the armpits and on the head. This makes us look very different when seen at a distance, and even more so when viewed close up.

The naked ape

Nakedness has become a species flag. In primeval times it was an identification signal that was immediately visible when another animal encountered a primitive human being. This role as a visual label may be its primary function. Its oddity is compounded by the development of excessively lengthy head hair and, in the males, of a long, shaggy beard. Imagine what a group of primeval humans must have looked like before there were clothes, razors or hairdressers. Standing up on their hind legs, with a massive expanse of head hair looming over stark naked bodies, they must have appeared frighteningly different from any of the other apes or monkeys.

There are those who have argued that humans are as hairy as their close relatives because there are just as many individual hairs on our bodies as on those of the other apes. It is true that evolution has not removed all trace of human body hairs, but it is obvious enough that the tiny, stunted hairs that have survived as reminders of what was once there in abundance no longer provide a functional coat of thick fur. Our naked skin may also have had other advantages. Losing a heavy coat of fur and developing plentiful sweat glands could be a great help in avoiding overheating in a hot environment. When our ancestors started to hunt and engage in prolonged chases this would have been of considerable assistance.

Walking on two legs instead of four freed up the front feet and allowed them to evolve into specialized hands.

It has been suggested that our ancestors spent time in the water, wading and swimming – the loss of a fur coat would make them more streamlined. It could also help to reduce infestations of skin parasites. All of these factors may have contributed to human nakedness and should be seen as a collection of possible advantages, working together in the same evolutionary direction.

Human beings are the only truly bipedal primate. Other species may rear up on their hind legs occasionally for brief periods of time, but none of them is a long-distance walker, or runner. It was this major change in locomotion that started the human species on its long success story, and again, several factors were at work. Walking on two legs instead of four freed up the front feet and allowed them to evolve into specialized hands. These became capable of a precision grip for fine work and a more powerful grip for heavy jobs, such as lifting. They also enabled our species to throw, grasp, punch, stroke, fondle, embrace, gesture, and a hundred other manual activities, the kind of actions that made early human societies so successful.

Walking tall

Apart from freeing the hands, bipedal locomotion also helped our early ancestors to see over tall grasses and other obstacles, and to scan the horizon. When we took to the water, being able to

Two legs better

Bipedality – walking on two feet – may have made early humans look remarkably tall and intimidating to quadrupedal (four-legged) rivals. The standing human body shows almost a straight line from the base of the skull down through the spinal column and the pelvis (hip bone) to the legs and feet.

Skull and spine

In the human skull the joint with the neck and the hole through which the spinal cord passes (foramen magnum) are in the middle of the skull's base, balancing the head over the spine. In the other apes the joint is towards the lower rear of the skull, so the spinal column joins at an angle.

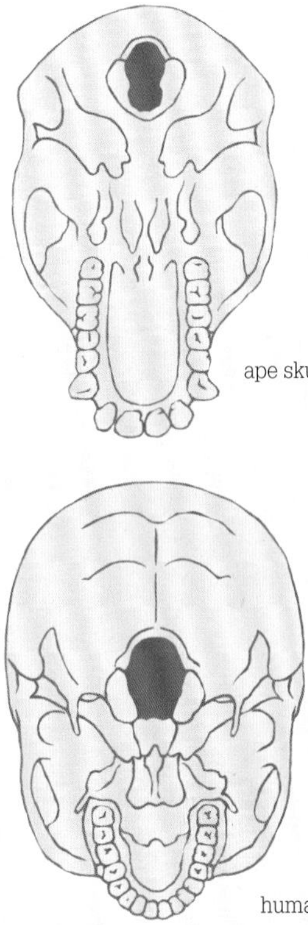

ape skull

human skull

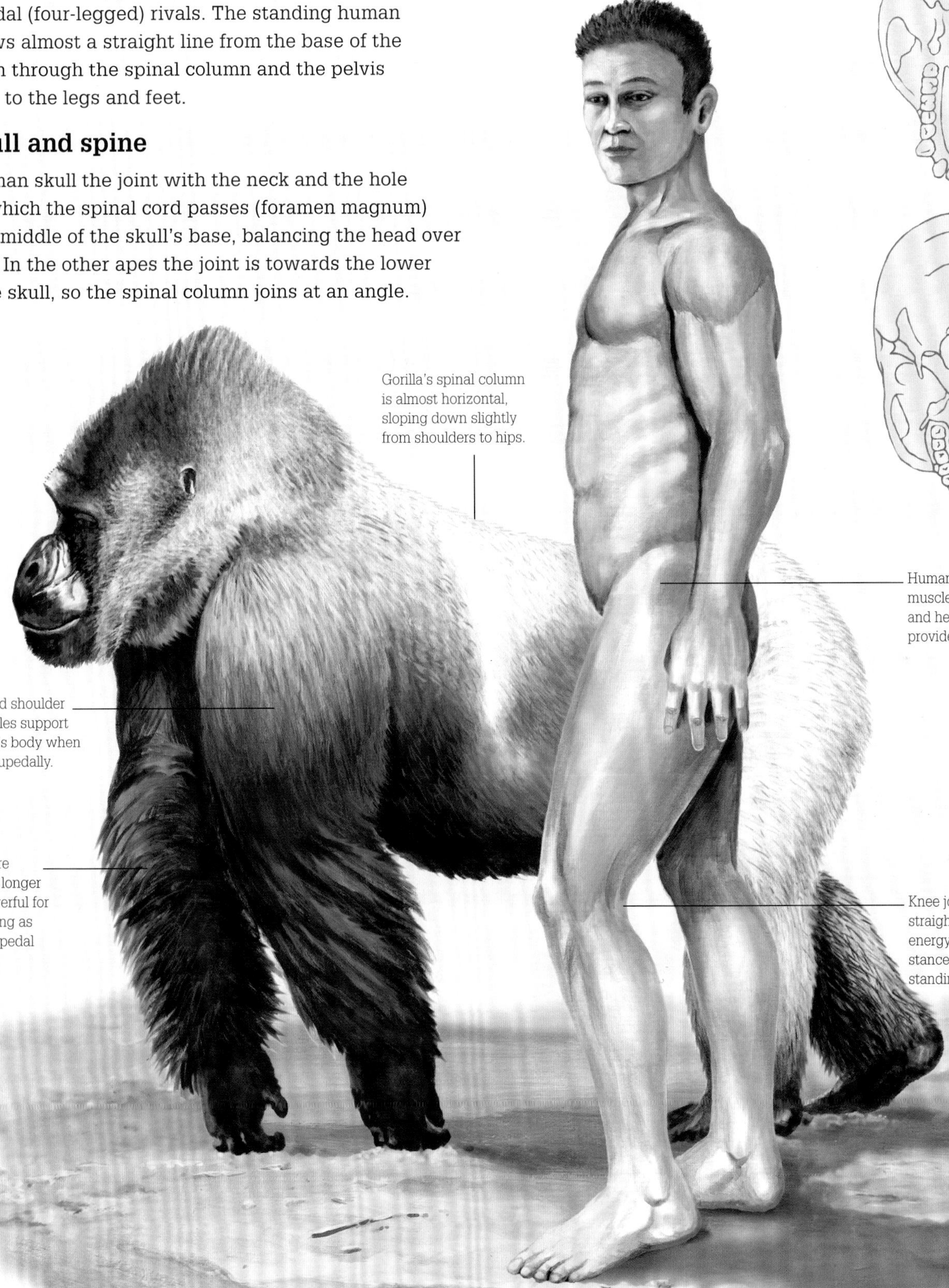

Gorilla's spinal column is almost horizontal, sloping down slightly from shoulders to hips.

Human's large hip and thigh muscles help to balance torso and head over legs and provide thrust for walking.

Well-developed shoulder and arm muscles support front of gorilla's body when moving quadrupedally.

Gorilla arms are proportionally longer and more powerful for knuckle-walking as part of quadrupedal gait.

Knee joints can 'lock' straight to save energy for a stable stance when standing.

stand on our hind legs with ease would again have been of great assistance. There may also have been an advantage when dealing with predators. A vertical posture would have made us appear larger and more threatening than we actually are, especially if accompanied by raised arms and loud human roars. There is also evidence that predator attacks may be triggered by a moving horizontal silhouette, since most of their prey are four-footed ungulates. If an ape is going to come down out of the trees and compete with ground-level predators, it has to look as little like an antelope as possible. The sight of a strangely vertical body might cause confusion in the mind of a big predator and give the early humans yet another advantage.

The development of language

A major threshold was passed when the walking ape first began to communicate, not just by grunts and screams like other apes, but by words. Verbal communication probably began as a way of identifying predators and food. With a few exceptions, relating to the identification of predators, the non-verbal vocalizations of other apes and monkeys are all about emotional states. They say things like 'I am angry', 'I am amorous', or 'I am afraid'. They tell about the mood of the moment, but they say nothing about specific objects. If an early man wished to communicate about a fish, or a bird, he made a sound that was only used in connection with a fish or a bird. If he added a verb, saying 'catch fish' or 'kill bird', for instance, he could then start to indicate his intention. The next stage was the development of a simple grammar that allowed humans both to describe the past and to plan future action.

The development of language formed the basis for even more impressive human development. However, yet another something extra was needed for us to become the all-conquering species that we now are.

The child in us all

With the ability to use verbal communication to plan future events, our ancient ancestors were able to develop hunting tactics and strategies. They were able to organize their small communities more efficiently. But in order to do this they needed extra mental flexibility and imagination. They achieved this in a strange way, by becoming more juvenile. All animals play energetically when they are young, but they give this up when they become adult. By becoming 'juvenile apes', even when fully grown, early humans were able to extend childhood curiosity and inquisitiveness right through into adulthood. The walking, talking ape became a creature capable of great feats of invention. By applying the eager, exploratory attitude of childhood to adult projects, primeval humans were able to develop new technologies and to explore new environments. As successful hunter-gatherers, they began to spread out of Africa and around the globe at an unprecedented rate.

By becoming 'juvenile apes', even when fully grown, early humans were able to extend childhood curiosity and inquisitiveness right through into adulthood.

The power of thought

With all the new possibilities that early humans encountered as they spread, their brains began to grow more complex. There was so much now to do, to say, to invent, to explore, to build, that only those individuals with increasingly complicated brains inside their skulls would be able to thrive. In a relatively short space of time, evolutionarily speaking, the human brain became three times the size of those of other apes. However, there was more than just an increase in quantity: there was an increase in quality as well. So it was that the human species became the most intelligent animal on Earth and, as the millennia flew past, came eventually to dominate the entire planet. Although emotionally still very much an ape, awash with all those basic apelike feelings of envy, jealousy, pride, greed, fear, anger, and the rest, this new species was something very special and very different, truly the most extraordinary animal ever to have existed. By a series of almost accidental steps, the primates had given rise to a monster, a splendid, brilliant monster, but also one that would soon threaten the existence of all of its closest relatives and many more species besides.

Exploring the world Young apes like this thoughtful orangutan have an enormous curiosity and capacity for exploring their environment. This usually fades as they age, but the human ape has retained its exploratory and inventive attitudes as an adult.

How the human species advanced

When our ancient ancestors left the trees, came down to the ground, and increased their intake of animal food, they faced dramatic new challenges. These altered their whole way of life, making them socially very different from the other apes. Living in small tribes, as hunter-gatherers, they had to evolve a new kind of personality if they were to thrive. And thrive they did, quickly spreading out from their African homeland to conquer Europe, Asia, Australia and the Americas. In no time at all, biologically speaking, the tribal ape was to become urban man.

When the earliest humans became ground-dwelling hunters they had to adapt rapidly to the new demands this put upon them. Taking small animal prey, fish, shellfish, insects, eggs and the like was not a problem. Finding and eating such items was little different from finding fruits, nuts, seeds, or leaves. However, when the first hunters turned their attention to bigger prey, they needed all the help they could get. Cooperative hunting had to develop as a new pattern of human behaviour. Furthermore, the hunters had to go off in search of their new prey, sometimes on long trips, leaving the females and their young behind. A division of labour began to develop, with males becoming specialized as risk-taking athletes. They were up against the already highly refined predators – animals with immense physical power, and sharp claws and teeth. Human hunters had to resort to brain rather then brawn and to active cooperation involving clever strategies and careful planning. They had two choices. By ganging up into a frightening group, they could either drive other predators off their kills and scavenge the meat or they could make their own kills. Being adaptable, they probably did both.

Confronting our fears

Competing with large carnivores must have been a daunting prospect. In the past, the typical reaction of primates to meeting animals like lions and leopards was to scamper up into the safety of the treetops, but our early ancestors had to stand their ground. This required the evolution of a new personality trait – courage. The hunters had to become brave and take serious risks. The development of spears and other weapons made hunts increasingly efficient and productive, and the huge boost that the large prey brought to the human diet made a major contribution to the early success of the small tribes. It also brought about a change in the digestive system, with the enzymes needed for breaking down meat becoming more advanced.

Our early ancestors had to stand their ground. This required the evolution of a new personality trait – courage.

Because males and females had to split up when hunts were in progress, it became increasingly important for there to be a fixed settlement to which the hunters could return. Nests of the type used by other apes were not permanent or protective enough. Something more was needed, and so architecture was born. For hundreds of thousands of years this consisted of no more than little round huts of the simplest kind, but they were enough. They gave hunting apes a home base, a specific location to which they could return.

The oldest known trace of a round hut dates from 1,900,000 years ago in the Olduvai Gorge in Tanzania. Much later, a mere 300,000 years ago, there was a cluster of similar huts on what is now the French Riviera. Remains of this settlement were discovered recently, revealing that the huts had been built and rebuilt repeatedly and therefore that the settlement was occupied over a long period. Food remains revealed that the occupants of these huts were dining on oysters, venison and wild boar and that they fashioned flint tools while sitting around the hearth in the middle of each of the huts. That type of small settlement was to last for a very long time. Indeed, in the most remote tribal communities of the present day, similar places can still be found.

Past and future
In the 1950s ITV's *Zoo Time* became the UK's first weekly wildlife series aimed at children. Today's efforts at saving our closest living cousins should likewise inspire conservationists of the future.

Agriculture and towns

Where hunter-gatherers were most successful, their tiny settlements started to grow into villages. Then, about 10,000 years ago, a major new threshold was passed. Instead of simply gathering wild fruits, seeds, and roots, certain villages started to cultivate them and plant them as crops. When wild animals came to raid these new crops, the villagers caught them, tamed them and started to breed them for food. Farming had begun and soon a new luxury appeared – the stored food surplus. Before long the villages grew into market towns, where produce could be exchanged. A merchant class appeared and eventually specialist craftsmen. Over the next few thousand years, towns began to swell into cities and civilizations sprang up. Today cities are still occupied by people who have not yet had the time to evolve into a new urban species. The urban ape is still a tribal animal at heart, with the mentality of a hunter-gatherer. It says a great deal about the flexibility of the human species that we can survive the levels of overcrowding that we now endure. We manage this new situation in an ingenious way. As city-dwellers we continue to behave as if we are still members of a small tribe. Our address books list the members of our tribe, but it is a tribe that is interspersed with countless other small tribes, all interwoven into one huge, sprawling city population.

The urban ape is still a tribal animal at heart, with the mentality of a hunter-gatherer.

Whether, eventually, we will evolve to another, more advanced stage of human development is hard to say. As the world population increases, we will find ourselves put under more and more pressure as overcrowding grows. We will no doubt continue to cling to our tribal roots for as long as possible, if only because we rather enjoy the dramas that flow from the primitive emotions of love, hate, jealousy, loyalty, pride, and status that we share with our ape relatives. Whatever happens we will surely remain the most exploratory animal that has ever lived. This alone will almost certainly carry us on to greater successes and who knows, perhaps one day to other planets.

USM

The hunted ape

Wild apes as our enemies

It is hard to believe today, when so much effort is being expended to protect the few surviving wild apes, that not so long ago they were looked upon as savage monsters to be hunted and killed as trophies by intrepid Western explorers. There was no sympathy for them. They were greatly feared and universally viewed as dangerous beasts that would brutally tear you to pieces if you did not shoot them first. This attitude persisted until the first part of the 20th century.

In medieval Europe the great apes of Africa and beyond were unknown and the word ape was used for the tailless macaque known today as the Barbary ape. Barbary apes were roundly condemned for their arrogant attempts to imitate human beings. This gross impudence had led to them being labelled as the prototype for the imposter, the fraud, the hypocrite, and the flatterer. Early Christian fathers saw the 'ape' as a devilish counterfeit and an enemy of Christ. To them, it was a diabolical beast. An unholy fake, Satan himself became known as *Simia Dei*, or God's Ape. Martin Luther used the terms ape and Devil interchangeably.

When the wild great apes finally began to impose themselves on European symbolism in the 18th century, their impact was, inevitably, even more extreme. They came to symbolize the brutal rapist male, ready to carry any unsuspecting woman off to their vile lair and have their hideous way with her. The French naturalist Buffon reported that orangutans 'are passionately fond of women, who cannot pass through the woods which they inhabit, without these animals immediately attacking and ravishing them'. An Italian travelling in Borneo in the 19th century noted that 'The Dyaks tell many a tale about women being carried off by orangutans. No doubt the thing in itself is possible, for an adult [orangutan] is certainly strong enough to carry off a woman'. In America, even Thomas Jefferson, with his enlightened approach to the immorality of slavery, commented on 'the preference of the Oran-utan [sic] for the black woman over those of his own species'.

Long live Kong! This giant gorilla, with his tragically doomed love for the human heroine, receives a Hollywood spruce-up every decade or two, as well as making guest appearances in other movie hokum spectaculars.

The vicious king of Darkest Africa

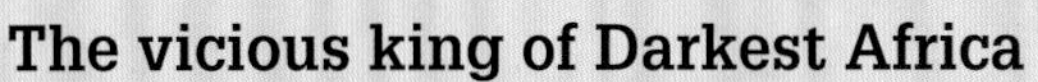

In the 19th century, when the mighty gorilla was first discovered in Africa, this new ape villain stole the orangutan's thunder. For years there had been tales of a legendary, abominable Man-of-the-Woods, a terrible hairy monster living in the dense undergrowth of the tropical forests of West Africa. Now at last it was a scientific reality. The American ape-hunter Paul du Chaillu, who claimed to be the first white man to hunt down gorillas in the wild, vividly described their horrific appearance, ferocity and malignity. He did, however, issue a denial that they carried off women from the native villages. His denial, spoiling a good story that had become deeply entrenched in Western thinking, was ignored, and as late as the 1880s, a medal-winning bronze sculpture entitled *Gorilla and Woman* was exhibited in Paris. It showed the classic image of a giant wild ape, roaring defiance and carrying off a helpless, struggling woman.

This famous sculpture may well have influenced the notorious scene in the film *King Kong* (1932), when Kong carries off the screaming heroine in his giant paw. This early 20th century tale still has the gorilla as a kidnapper of women and a symbol of male sexual brutality, but something new is added. With our attractive blonde heroine he weakens, and although he bites the heads off sundry natives, he finds her strangely appealing and does her no harm.

◀◀ Contrasting technologies
(Previous page)
Orangutans in the wild are known to use simple tools such as sticks to prise open prickly fruit. But in captivity they will readily copy human behaviour and manipulate far more complex devices, including turning the lens on the photographer.

▶ Changing attitude
By the 1920s gorilla-hunting was in full swing as tales of this ape's savage rage spread.

Glorified slaughter

King Kong was only fiction, but in real life the gorillas of West Africa were already being subjected to a ruthless onslaught from an invasion of trigger-happy big game sportsmen. R.M. Ballantyne's adventure story *The Gorilla Hunters* (1861) epitomizes the Victorian attitude towards the greatest of the apes. Here is his description of a close encounter with an adult male: 'The bushes in front of us were torn aside and the most horrible monster I ever saw…stood before us… His jet black visage was working with an expression of rage that was fearfully satanic. His eyes glared horribly… Peterkin fired, and the gorilla dropped like a stone, uttering a heavy groan. What a monstrous brute!'

This immensely popular book was reprinted many times, right up to the 1930s. Appearing during this time-span were many photographs taken of freshly shot gorillas, their huge carcasses awkwardly propped up for the triumphant hunter to impress the folk back home. For countless millennia the gentle gorillas had been minding their own business, but this was the beginning of the end for them.

The gorillas were being subjected to a ruthless onslaught from an invasion of trigger-happy big game sportsmen.

Sadly, the arrival of gorilla protection has not entirely stamped out the existence of gorilla hunters. The Westerners who shot gorillas for sport may have vanished but another breed of hunter is now on the prowl: locals who shoot gorillas mainly for their meat. It has been reported that, in Cameroon, for example, 'gorilla hands and feet are considered a delicacy and are served to the guest of honor at official functions'. With customs like that surviving into the 21st century, the gorillas are going to find themselves staring down the barrel of a gun for quite a while. The hunted ape, tragically, has yet to fade into history.

The performing ape

Our cousins as entertainers

Me Cheeta, Him Tarzan
Lex Barker took the lead role in *Tarzan and the She-Devil* (1953), or was it his lively sidekick, Cheeta, allegedly played in many Tarzan movies by a chimpanzee called Jiggs?

While adult apes were being hunted in the wild by Victorian adventurers, a few young ones found their way to Europe, where they provoked curiosity rather than fear. They became entertainers, presented to the public to provide amusement with their antics. At first, these performing apes were seen only in zoos, but as the years passed they began appearing in circuses, on the stage, in films, and eventually on television. The era of the ape clown was upon us.

One of the first live apes to be shown to the public was a small male chimpanzee called Tommy. The ship bringing him from Africa docked at Bristol late in 1835 and London Zoo sent one of their keepers to collect him and bring him to the zoo in a stagecoach. An excited public was astonished at how human he seemed to be. A poet of the day wrote: 'The folk in town are nearly wild to go and see the monkey-child, in Garden of Zoology whose proper name is Chimpanzee. To keep this baby free from hurt, he's dressed in a cap and a Guernsey shirt; they've got him a nurse, and he sits on her knee, and she calls him her Tommy Chimpanzee'.

Sadly, the nurse failed to keep Tommy in good condition and he was dead within six months. A few years later, in 1837, the zoo received its first orangutan, a little female aged about three years, called Jenny. She was also dressed up like a small child and was treated, it was claimed, like 'an Asiatic Princess'. Although, like Tommy, she would not live long, she was destined to play a major role in the development of human thought, for, on 28 March 1838, a young visitor to the zoo was stunned by his encounter and drew a momentous conclusion from it.

The young man in question was Charles Darwin, freshly back from his epic voyage on the *Beagle*, his head buzzing with ideas about animal evolution. He had never seen a living ape before and was so fascinated by the behaviour of the orangutan that he wrote in his notebook: 'Let man visit Ourang-outang in domestication…see its intelligence… see its affection…see its passion…and let him dare to boast of his proud pre-eminence'. Significantly, he also wrote: 'Man in his arrogance thinks himself a great work, worthy the interposition of a deity. More humble and I believe true to consider him created from animals'. It would be over 20 years before Darwin's theory of evolution would be published in *On the Origin of Species*, but already at this much earlier date a little ape had helped clear his mind.

Before the arrival of living great apes in Europe, mankind could comfortably see itself as totally above and separate from the rest of the animal world.

A meeting with the Queen

When she died prematurely Jenny was quietly replaced with a second orangutan, also called Jenny. Such was the fame of this one that, in 1842, she took tea with Queen Victoria and Prince Albert. The Queen, it seems, was not amused, remarking afterwards that she found the ape 'frightful and painfully and disagreeably human'.

What Victoria had anticipated with distaste, of course, was the fact that apes and humans would eventually be seen as truly related, rather than simply as comically similar. Before the arrival of living great apes in Europe, mankind could comfortably see itself as totally above and separate from the rest of the animal world. Now a disturbing bridge was being built between the two worlds, and silent anxieties were being felt about the validity of religious dogma.

Whatever some people might have thought, right through the 19th century the majority were happy enough to go on laughing at the zoo apes and their quaint ways, and zoos persisted in

See my agent
Future US President Ronald Reagan was once better known for his B-movie career, especially *Bedtime for Bonzo* (1951) with Diana Lynn.
Several orangutans starred as Clyde alongside Clint Eastwood in the 'Any Which Way' movies of the 1970s–80s.
Chimpanzee 'tea parties' were a popular laugh-at attraction in some regions until the 1990s.
In the 1950s the chimpanzee J Fred Muggs spent more than four years as co-star, with Dave Garroway, on NBC's long-running *Today* show.

dressing up their apes as parodies of human personalities. One of the most famous of all these apes was a young male called Consul who held sway at Manchester Zoo in the 1890s. Consul was remarkable for his wardrobe. Instead of dressing him like an infant, the zoo gave him elaborate adult clothing, a fancy hat, smart boots, a walking stick, and a pipe.

The chimps' tea party

In the early 20th century animal collectors were out in force, shooting adult female apes and taking their tiny babies. Bottle-fed, the babies often did well and soon the zoos of the world were all proudly displaying their young apes. Some of them were kept in proper enclosures for breeding, but others were still treated as human caricatures. As late as the 1950s, London Zoo was still staging its famous chimpanzee tea-party every summer afternoon, enjoyed by huge crowds of laughing children.

At the cinema, Tarzan was now a popular figure, aided and abetted by his famous chimpanzee called Cheeta, a resourceful animal who was at least allowed to be intelligent as well as funny. With the advent of television after World War II, new ape stars began to appear. In the 1950s, I myself was persuaded to appear each week with a young male chimpanzee called Congo. As a zoologist I refused to follow the funny-little-ape tradition and insisted, against much opposition, that Congo would appear unclothed. I also insisted that his weekly appearances should always be designed to display his amazing intelligence, rather than his ability to copy human behaviour. It was, however, an uphill struggle. Congo's ability to wreck carefully designed tests and cause me endless trouble made him into a much-loved star. Unsolicited gifts for him arrived by the sackful. When he caught a cold one week and was confined to bed he received over 20,000 Get Well cards and letters.

In the United States a chimpanzee called J Fred Muggs was also making his name as a TV celebrity, but he was still at the dressed-up stage, and little scientific enquiry was involved in his appearances. He was always immaculately dressed and at one point was filmed making a worldwide tour, staying at the best hotels and being driven in a large limousine. However, this was almost the beginning of the end for the performing apes. They might linger on in TV commercials and as actors in the occasional, old-fashioned Hollywood movie, but as the 1950s drew to a close they were about to be replaced by something much more informative, the real, wild ape, observed in detail in its natural habitat for the first time.

As late as the 1950s, London Zoo was still staging its famous chimpanzee tea-party every summer afternoon.

Apes observed

Changing attitudes

As early as the 19th century, a few intrepid explorers had expressed the desire not to kill wild apes but to observe them in their homeland and learn how they lived. They had little success. During the first half of the 20th century the emphasis shifted to studies in captivity. Great primate centres were established in the United States and work began on testing the abilities and intelligence of the non-human apes, almost always using chimpanzees. A great deal was learned, but in a sense, the psychologists involved were putting the cart before the horse. Only with the arrival of a new breed of field worker in the 1960s would this situation finally change. When that happened, our understanding of the great apes would take a huge leap forward.

One of the first people to attempt to observe apes in their natural habitat was the American naturalist R L Garner. He had spent a great deal of time recording the sounds of captive primates but wanted to hear how they communicated with one another in the wild. To achieve his goal he spent two and a half years in the tropical forests of the French Congo, a courageous undertaking in the 1890s. He learned the cries of the apes and began communicating with them. On one occasion he was so successful that a large male chimpanzee bounded up to him. At this point, fearing for his life, Garner reverted to the Victorian norm and shot the animal. Although Garner learned something about the language of wild great apes, he discovered little about their lifestyle. This was because he spent most of his time inside a large wire cage he had built to protect himself from ape attacks. He wrote a book about his 'observations', but a careful reading reveals that almost all his comments are based on stories told to him by the local people. He tells how 'a gorilla king, called *ikomba njina* by the natives, along with his wives and their young together make up a nomadic family of about ten to twelve individuals'. Here he presents a more accurate picture of a gorilla group than his predecessors, but he spoils it by adding that, while the king gorilla sits in the shade of a tree, his family busily collect food for him and bring it to where he is regally resting. They then, he says, present it to their lord and master.

Nissen was able to refute the popular belief that chimpanzees lived in family units rather like humans.

New discoveries

Almost 30 years would elapse following Garner's work before there would be another serious attempt to study apes in the wild. This time it was an American psychologist called Henry Nissen who undertook the challenge. He set off to French Guinea to observe chimpanzees in their natural environment and then bring some of them back to the United States for further studies in captivity. In 1930 he spent three months tracking and watching chimpanzees, and managed to make observations on a total of 49 days. At this point in history nobody had seen more of wild chimpanzees than him, and his observations were therefore of great interest.

Nissen was able to refute the popular belief that chimpanzees lived in family units rather like humans. He saw that there were often several adult males in a group and that one group would meet up with another, merge and then separate again. Groups were seen to roam around a home region, but without a fixed territory. Nissen also studied their bed-making activities, their cries, their drumming, and their care of the young, but one overall observation made a lasting impression on him. He was astonished at how lively and extroverted the wild chimpanzees were. He was used to seeing bored, introverted, neurotic apes living in small bare cages in

Captive audience
In the 1940s, the great apes attracted huge crowds to zoos like this one in Central Park, New York. Over the following decades zoologists and the public alike became aware that captive animals could be healthy and happy only in conditions that were more akin to their natural habitat than this barren cage.

captivity, and the busy, playful groups he was confronted with in Africa gave him an entirely different attitude to the chimpanzee as a species. Unfortunately, it did not stop him collecting 16 apes and condemning them to the monotony of a captive existence back in the United States.

A more direct approach

In the years that followed there were other sketchy reports on the behaviour of great apes in the wild, but it was not until 1959 that the first in-depth observational attempt was undertaken. The pioneer who began the whole modern trend for prolonged, close observational studies of apes in the wild was the American zoologist George Schaller, arguably the greatest field naturalist of the 20th century. He spent two years tracking mountain gorillas on the slopes of the Virunga volcanoes in Central Africa. The new approach he adopted was to pave the way for a major breakthrough in our understanding of all the great apes. His method was brave, because nobody had dared try it before, and it was simple. Instead of standing inside a cage like Garner, camouflaging himself, or hiding in the undergrowth, he made himself obvious to the animals. He would climb uphill slowly and in full view of a group of gorillas, settle himself quietly on a tree stump or a low branch, and then pretend to ignore the apes. After a while they would relax and some of them would approach to within a couple of metres to examine him more closely. The more he did this, the more he was accepted, and he was never once attacked throughout his two-year study.

Instead of standing inside a cage, camouflaging himself, or hiding in the undergrowth, Schaller made himself obvious to the animals.

Unlike his predecessors, he was able to make lengthy observations of the family life of the gorillas. He encountered them no fewer than 314 times and scored a total of 466 hours of direct observation of them, usually from a distance of no more than 50 metres (165 feet). His golden rules for successful gorilla-watching were summed up as follows: carry no firearms, move slowly, minimize gestures, remain alone, wear the same clothes, and wear drab green colours.

Apes studied in the wild

The 1960s witnessed a revolution in the field studies of great apes. And, surprisingly, it came, not from eminent, experienced zoologists, but from recklessly brave young women. While George Schaller was in the field with his gorillas, an Englishwoman called Jane Goodall was preparing to make a similar investigation into the behaviour of wild chimpanzees. Although she had no formal scientific training, Jane Goodall had caught the eye of the famous African fossil-hunter, Louis Leakey, who arranged funds for her to undertake a long study of the apes living at Tanzania's Gombe Stream Chimpanzee Reserve. Little did anyone realize at the time that this would turn out to be the most famous and most influential of all wild ape projects.

Arriving at Gombe in July 1960, Jane took the unprecedented step of making physical contact with the wild chimpanzees. Those of us who knew her at the time were deeply concerned for her safety. Chimpanzees are not shy, gentle giants like gorillas. They are boisterous, volatile and, despite their smaller size, immensely powerful. Jane knew this, but ignored it. Leakey had told her that, if she kept calm and meant them no harm, the animals would sense this and she would not be attacked. She followed her apes day after day, getting closer and closer, until she was almost within touching distance. Barefoot and wearing only a shirt and shorts, she was incredibly vulnerable, and on many occasion was thumped by a male as she crouched silently on the ground. But eventually her patience won her the prize she wanted, and she was accepted into chimpanzee society.

Part of the family

Jane is the only person ever to have achieved this, and it gave her an unrivalled opportunity to make intimate observations of every aspect of chimpanzee social life. One of the disadvantages of being accepted in this way was that she also became involved in ape politics. Jane was targeted by a male chimpanzee, who would be aggressive if he saw her, hurling stones or knocking her over. She amassed so much information that she was established as the world's foremost expert on the subject, and in 1964 she set up the Gombe Stream Research Centre. Many pupils went to join in her studies there, and she herself remained at Gombe for about 20 years.

By the mid-1960s George Schaller had moved on to other projects and as a subject for field study the gorillas had become vacant. At this point a young American, called Dian Fossey, entered the scene. Like Jane, she had met Louis Leakey, and he had told her about Jane's achievements. Also like Jane, she was not an academic zoologist: she had been a therapist at a children's hospital in Kentucky. But the lure of Africa, and Louis Leakey's charismatic enthusiasm, won her over and she decided to try Jane's tactics with the mountain gorillas. George Schaller may have got close to them, but he had never attempted to make physical

A lifetime's dedication
Dian Fossey receives a friendly touch from Puck, a senior female in a mountain gorilla family she observed for many years. Puck was one of the most studied of all wild gorillas and died in May 2007, aged 38 years.

One of the group
Jane Goodall's unstinting commitment to primate studies, and especially her founding of the chimpanzee centre at Gombe, helped to alter public perceptions about how these great apes really lived. Posing for a family portrait in Gombe National Park, Tanzania. Trialling communication techniques with one of the sanctuary chimpanzees.

contact. Here was an exciting, if daunting, new challenge. Dian began her studies in 1966 and was soon as close to her apes as Jane had been to hers. Although the gorillas were much bigger and even more powerful, they were much more timid. Indeed, Dian was about to dispel, once and for all, the widely held belief that they were belligerent monsters. Like Jane, she became accepted as part of their family and soon images began to appear that showed the gorilla in its true light as a peaceful, retiring animal.

Regaining lost trust

That said, Dian did have occasional frightening moments. Before her arrival the only human contact some of the gorillas she encountered had experienced was when hunters were gunning down their families. Bearing that in mind, it is understandable that some of the dominant males initially became enraged, but treated with quiet respect, they soon returned the favour and before long Dian was amazing the scientific world with her detailed reports on gorilla life. She continued to work in Rwanda until 1985, when she was brutally murdered by someone who objected to her presence in the country. The problem she faced did not emanate from the gorillas. She was waging a constant war against poachers and was also unpopular with some of the local officials who wanted to promote tourism in the region, something she opposed because tourists brought human diseases to her gorillas. She also hated zoos and especially the animal collectors who tried to take young gorillas for them. So she had many enemies, all wishing her dead. Her killer was never identified. It was a tragic way for her to prove her point, that it is humans who are the savage monsters, not gorillas.

Dian Fossey dispelled, once and for all, the widely held belief that the gorillas were belligerent monsters.

Other landmark studies

In the latter part of the 20th century many more studies of wild great apes were undertaken, among them important investigations of chimpanzees by Frances and Vernon Reynolds in the Budongo Forest, and of orangutans by John Mackinnon and later by Birute Galdikas in Indonesia. As a result we have a much deeper understanding of the true personalities of the great apes and the way they spend their lives. This has helped enormously in destroying the old prejudices and in stimulating global efforts to save these species from extinction. As George Schaller had said, back in the early days of his pioneering field study, 'No one who looks into a gorilla's eyes can remain unchanged. The gap between ape and human vanishes…'

Ape intelligence

There is a tendency today to think of all captive great apes as abused, exploited and even tortured victims, but this overlooks the fact that many young apes have been kept as close companions, reared with considerable affection and greatly loved by their human carers. Those individuals that are hand-reared can prove to be invaluable ambassadors for their species. In particular, with sensitive handling, they can demonstrate the remarkable mental abilities and outstanding intelligence of the great apes, helping in the shift of public opinion, from laughing at apes to marvelling at them.

Some of the earliest intelligence tests using great apes were those carried out by the German psychologist Wolfgang Kohler in the Canary Islands during World War I. His basic test was to offer attractive food to chimpanzees, but in such a way that they could not obtain it directly, but had to work out a way to reach it. In one case, the food was clearly visible and close by, but separated from the apes by a wide wire barrier. To obtain it they had to walk around the barrier. This meant that the first steps they took to obtain their goal would be taking them away from the food. This detour solution was beyond the ability of other animals tested, but the apes found it easy. They also solved the problem of how to pull food towards them if it was out of reach but tied to the end of a long piece of string, or how to employ a stick to scrape food nearer and nearer until they could reach it.

Trusted friend
(Previous page)
Jane Goodall's 40-plus years of involvement with chimpanzees has led to a multitude of observations on the intimacies of their personal lives. In 1977 she founded the Jane Goodall Institute, whose aims include conservation of primate habitat and expanding non-invasive research on chimpanzees.

Kohler was deeply impressed by the way his chimpanzees reacted in these situations. In the past it had been argued that animal intelligence was always based on trial and error, but Kohler pointed out that, in the case of his chimpanzees, something much more 'human' was taking place. After the apes had tried and failed the obvious, direct way of getting the food, they would suddenly see a more indirect way of tackling the problem and try that. Kohler called this the 'aha moment'. Later, some researchers would call it the 'Kohler moment'. Officially it was referred to as insight learning and was quoted as an example of how close chimpanzees and humans are when it comes to solving simple intelligence tests.

Kohler's tests became more complicated, and the chimpanzees continued to surprise him. In one famous instance a chimpanzee called Sultan managed to fix two short sticks together to make a long stick with which he could reach his prize. This was more than tool-using, it was tool-making and many years later observations of wild chimpanzees revealed that they do indeed both use and make tools in the wild.

If a chimpanzee observes someone placing a coin in a slot machine, opening a drawer at the bottom of the machine and drawing out a packet of raisins, it will repeat the process without any hesitation

Grape expectations

In the 1950s I devised other tests at London Zoo. In these, the reward was always a single grape – a mere token. It was clear that the chimpanzees performing these tests enjoyed the fun of the test itself as much as the reward. In one test the grape had to be rolled down a zigzag slot. To make it roll, the chimpanzee had to push a stick into a hole on the left and press. The grape would then roll to the left and stop. The stick then had to be taken out of its hole and pushed into a hole on the right. This rolled the grape a little further and then it stopped again. Little by little, the grape zigzagged its way down a long slope until the ape could reach it and eat it. The chimpanzee never became impatient and understood the whole test after watching a single demonstration. In another experiment, a fairground coconut shy was scaled down to take grapes instead of nuts. This grape shy was to test the aiming ability of apes. The chimpanzee was offered a ball on the end of a chain and

Ape talking
A chimpanzee fingers the lexi-keys at Yerkes Language Research Center, near Atlanta, Georgia, USA.

had to swing it in an arc towards the grapes. It succeeded in knocking them down as well as any of the humans involved in the tests, and its level of intense concentration was almost painful to watch.

Chimpanzee versus human

One-trial learning is easy for a young chimpanzee. If it observes someone placing a coin in a slot machine, opening a drawer at the bottom of the machine, and drawing out a packet of raisins, it will, as soon as it is given a coin, repeat the process without any hesitation.

It is sometimes hard to credit just how clever and quick-witted chimpanzees can be. In some recent studies they were given a test in which they performed better than adult humans. The test was designed to examine skill in numerical recollection. Young chimps were first taught the order of numerals from 1 to 9. Then a chimpanzee or a human subject was placed in front of a computer screen on which the numbers 1,2,3,4,5,6,7,8, and 9 were displayed on nine small squares. The positions of the squares were randomized, as were the numbers on them. When the square with the number 1 on it was touched, the other eight all went blank. To succeed in the test the ape, or human, had to touch these other eight squares in the order of the numbers that used to be there. To the amazement of the experimenters, the chimpanzees were faster than the human students playing against them.

Even more extraordinary was a modified version of the same test in which numbers were flashed up on the screen for only a fifth of a second. The squares then all went blank and the subjects had to touch them in the correct numerical order. The speed of the test meant that there was no time to look around the screen. The random arrangement of numbers had to be memorized as a whole pattern at a single glance. In this test, five-year-old chimpanzees had a success rate of 80 per cent, compared with only 40 per cent for human adults. If anybody has ever harboured doubts about the intelligence of great apes, watching a film of these tests will dispel them.

Apes and sign language

The first investigators to attempt to communicate with a great ape by using sign language were Alan and Beatrix Gardner at the University of Nevada in 1967. Previous attempts to teach apes to develop simple vocal languages had always failed, but the Gardners had noticed that chimpanzees often use expressive hand movements in moments of excitement. The thought occurred to them that it might be possible to train a young chimpanzee called Washoe to make specific gestures related to particular objects or actions.

The Gardners watched for moments when Washoe would make a certain gesture. When she was tickled, for instance, she would bring her hands together in a protective action. The American Sign Language signal for 'more' is the bringing together of the fingertips of both hands. The Gardners used this to teach Washoe that if she wanted to invite more tickling she had to make a deliberate hands-together action. Once she had learned that this meant 'more', she was able to use it in other contexts, signalling more food or more drink. In other words, she had developed an abstract concept of 'more' and was able to generalize, a key element in language.

Later, once Washoe had become familiar with the idea of signing, the Gardners noticed that she could acquire new hand signals simply by observing them. This type of rapid learning would occur if they demonstrated to Washoe a link between, say, a banana and a particular hand movement, or the act of drinking and another movement. Another leap forward was made when Washoe started to combine signals. Having acquired separate gestures for food, drink, and open she astonished the Gardners by referring to their refrigerator, as open food drink, employing three single-unit signs to make a short sentence.

War of words

It was eventually announced that Washoe could use about 250 different hand signs, but critics felt that her language skills were being greatly exaggerated and that, in reality, she had acquired a vocabulary of only 20 or so words. The key to language is the ability to recombine and rearrange single units in a novel and constructive way, and it was suggested that there was much less of this inventiveness in Washoe's hand-signalling than was being claimed. One of the Gardners' critics, noted for his negativity, stated flatly that 'the argument that Washoe is the first non-human to acquire a human language is generally considered without scientific support'. True, Washoe was no Shakespeare, but it remains extremely important to see how far apes can go towards the beginnings of linguistic expression. They may be capable of no more than the very first step, but that in itself is a remarkable achievement and tells us something important about the complexity of the ape brain.

Washoe developed an abstract concept of 'more' and was able to generalize, a key element in language

In the 1970s Penny Patterson of Stanford University started a similar study to that of the Gardners, but instead of a chimpanzee, she worked with a young female lowland gorilla called Koko. The claims for Koko's sign language abilities outstripped those for Washoe. Koko, who was studied for over a quarter of a century, was said to use more than 1,000 hand signs and to understand more than 2,000 words of English.

The most startling feature of Koko's behaviour was said to be her ability to invent new signs to communicate novel thoughts: she is supposed to have combined signs for finger and bracelet to mean ring, and signs for water and bird to mean swan. Again, there were harsh criticisms, suggesting that the work lacked any kind of scientific rigour.

These attacks were rejected along the lines that the critics had a deep-seated fear of apes and humans being shown to be so close to one another. However, some of the criticisms did

The trials of Koko
The claimed abilities of lowland gorilla Koko (born 1971) have been central to studies of how great apes communicate. ◩ Koko replies to her chief human companion Francine 'Penny' Patterson (on the left) that yes, she is hungry. ◪ Koko asks for an orange by extending her left arm away from her body. ◪◪ Koko makes the sign for 'to listen', telling Penny she wants to listen to the phone, perhaps because she does not yet know the sign for telephone.

seem justified. One was that the gorilla often babbled with its hands, producing sequences of gestures out of which its teachers would select whatever they wanted in order to prove their point. But even allowing for the possibility of overenthusiastic interpretation, the achievement of apes like Koko cannot be ignored.

Voices of dissent

In response to the Gardners' and Penny Patterson's work, and also in the 1970s, a whole project was undertaken to undermine the reputations of Washoe and Koko. Called Project Nim, it was led by Herbert Terrace, a psychologist at Columbia University. His main objection was that apes lacked syntax, a key element in human language, but his findings must now take into account the discoveries of field workers in Nigeria in 2006 who found good evidence of a syntax-like natural communication system in the calls of wild *Cercopithecus* monkeys. If monkeys are capable of this, it should be easy enough for great apes.

In the 1990s, further support for ape sign language was provided by anthropologist Lyn Miles at the University of Tennessee. Miles claimed that a large male orangutan called Chantek learned about 150 words of sign language and invented signs of his own. He could generalize, using the gesture for Lyn to apply to all carers, but never to strangers. And he occasionally lied, saying, for example, that he wanted to urinate when in reality he only wanted the fun of playing with the knobs in the bathroom.

Also in the 1990s, a new type of ape language study was begun by Emily Savage-Rumbaugh at Georgia State University, using a bonobo called Kanzi. Instead of making gestures, Kanzi was taught to point at printed symbols on a keyboard. When shown over 600 sentences, making requests employing novel combinations of words, Kanzi responded correctly in about three-quarters of the cases. If these claims are justified, they put Kanzi on a par with a human child in its third year.

Miles claimed that Chantek learned about 150 words of sign language and invented signs of his own.

It is sad that these ape sign-language studies have polarized research workers into two warring groups, one claiming too much and the other claiming too little. The fact that apes are capable of expressing even the tiniest germ of language is fascinating enough, and it can only be hoped that the arguments of the extreme proponents and extreme opponents of past years will eventually give way to a more balanced, objective assessment in the future.

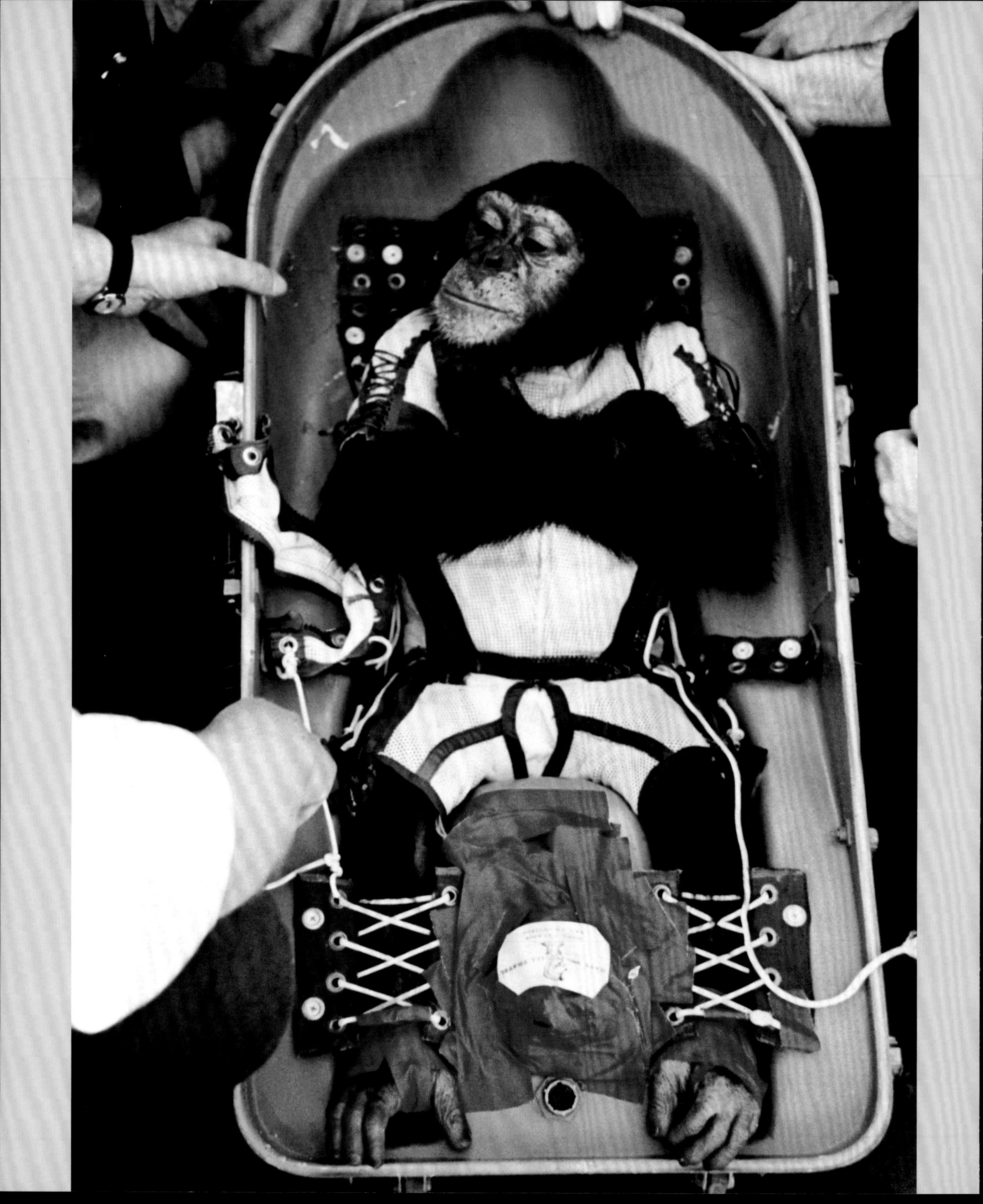

Apes in experiments

Being used as surrogate humans

The use of great apes, especially chimpanzees, in the fields of medical research, psychological testing, and even space exploration has been an uncomfortable chapter in our relationship with our nearest relatives. The great irony of the situation is that it is precisely because they are so closely related to us that they are considered such useful laboratory subjects. In other words, the very reason that they should not be subjected to these experimental procedures is also the reason that they are chosen for them. In recent years there has been an increasingly vociferous opposition to this type of research work, but in too many places it still continues.

The use of chimpanzees as captive research animals began in earnest in the 1920s, but at that date the focus of the investigations was chimpanzee behaviour and intelligence. In the 1940s, the direction of the research began to shift more towards invasive, biomedical experiments, but it was not until the early 1960s that biomedical research with apes was to become the subject of a major project. This project involved the establishment of seven large, government-funded Regional Primate Research Centers in the United States, the first of which opened in 1962. A primary goal of these new establishments was the improvement of human health, using results obtained from experiments performed on non-human primates. Medical areas covered included 'cardiovascular, neurological, reproductive, infectious disease, vision, aging, and diseases affecting women's health'.

Back in the 1950s, the American Air Force imported 65 wild-caught chimpanzees to set up a colony of potential astro-chimps or chimponauts

By the end of the 20th century, the seven American institutions held a total of 18,300 non-human primates of 32 species and employed 1,200 research scientists. The majority of these primates were monkeys, but over 1,800 of them were chimpanzees. Gorillas and orangutans had always managed to escape 'medical service'. By the year 2007 the chimpanzee figure had fallen to about 1,300, largely due to the failure of a major investigation into the AIDS virus, when it was found that chimpanzees exposed to HIV do not progress to AIDS. As the figures for chimpanzees being used in research declined, more and more of the surplus animals were sent to live in sanctuaries though many still remain in dark isolated cages.

To boldly go...

In addition to its use of apes in biomedical research, the United States also enlisted the help of chimpanzees in its conquest of space. Back in the 1950s, the American Air Force imported 65 wild-caught chimpanzees to set up a colony of potential astro-chimps or chimponauts. From these, a group of flight candidates was whittled down to eighteen and finally just six. One of these was a young male

Space pioneer The USA's first ape in space, chimpanzee Ham (1957–1983) made his historic 16-minute trip in 1961. His skills at trained tasks such as pushing levers were monitored, to show that his abilities were not affected by weightlessness and other conditions.

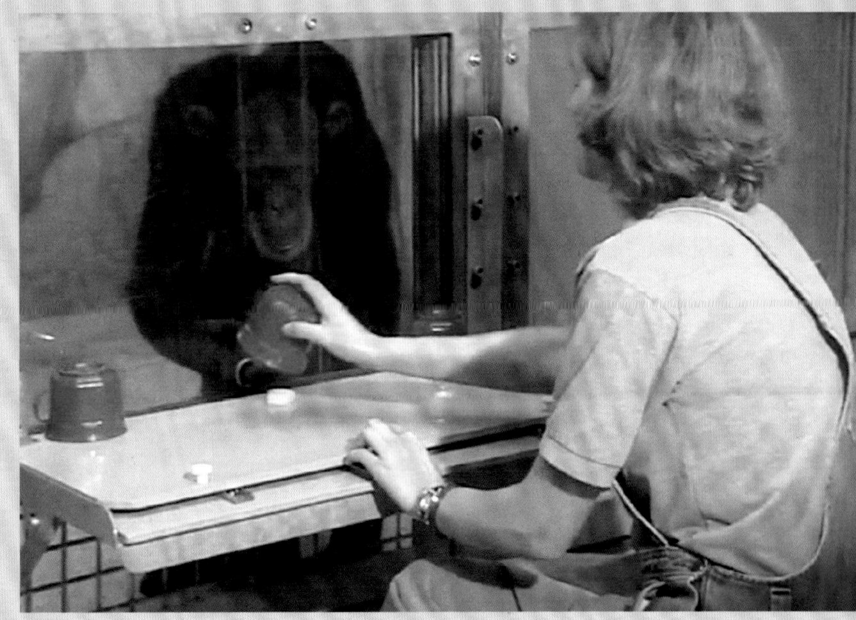

Now you see it... A chimpanzee correctly identifies the cup under which a researcher has hidden a banana. In this sort of spatial task chimps have been shown to outperform two-and-a half-year-old humans.

called Ham, whose name was an acronym for Holloman Aerospace Medical-center, and who, at the age of four, was destined to make history by becoming the first ape in space.

It was reported that 'in his pre-flight training, Ham was taught to push a lever within five seconds of seeing a flashing blue light; failure to do so would result in an application of positive punishment in the form of a mild electric shock to the soles of his feet, while a correct response earned him a banana pellet'. When fully trained, he was launched into outer space in January 1961. There, to everyone's delight, 'Ham's lever-pushing performance was only a fraction of a second slower than on Earth, demonstrating that tasks could be performed in space'.

Ham's flight lasted only 16 minutes, but 10 months later, a second chimpanzee called Enos successfully orbited the Earth. Other Air Force chimpanzees, failing to make it into space, were subjected to a variety of tests related to the astronaut programme, including impact, weightlessness and sleep-deprivation studies. Eventually the Air Force lost interest in their ape project, and the surviving apes – by then there were 143 of them – were disposed of to live out their lives in sanctuaries.

Growing opposition

In recent years there has been more and more opposition to the use of great apes in experimental research. The field studies of wild apes in their natural habitats have provided insights into their way of life that has made them appear so close to humanity that the idea of treating them as laboratory specimens has become increasingly offensive. Many countries have already placed a formal ban on any kind of experimenting with great apes. In 1997 Britain announced that it would no longer issue licences for research on great apes. In 2000 New Zealand adopted similar measures and in 2002 The Netherlands followed suit. In 2003 Australia restricted the use of great apes and Sweden banned research on all apes, gibbons included. In 2006 Austria, Switzerland, and Japan also fell into line.

The idea of treating great apes as laboratory specimens has become increasingly offensive.

In 2007 a study of 749 scientific papers published on biomedical research using chimpanzees as subjects came to a damning conclusion. The authors summed up by stating that their analysis 'demonstrated that there was no essential or even a significant contribution to the development of human treatment through the use of chimpanzees'. Little wonder then that the leaders of animal welfare movements have become so vociferous about the future of great apes in captivity.

Today, those welfare groups that are opposed to the keeping of great apes in captivity are making sweeping demands. In a recent Declaration on Great Apes they have insisted on the introduction of three basic rights for all chimpanzees, gorillas, and orangutans. These are: the right of life; the protection of individual liberty; the prohibition of torture.

If obeyed, these demands would mean the end to all experiments with great apes and a total ban on the capture of any wild ape in the future. It would also mean the return to the wild state of all great apes at present in captivity, both in research laboratories and in zoos. Although this would certainly put an end to the abuse and exploitation of great apes in those countries agreeing to this Bill of Ape Rights, in its idealism it has several shortcomings, as has been discussed elsewhere. A more practical – if less idealistic – suggestion would be to introduce a moratorium on the capture of any further wild apes. That and a worldwide ban on biomedical research using great apes as laboratory specimens would go a long way towards their long-term protection.

Planet of the apes?
Charlton Heston (centre) stands before the judicial council of orangutans in a still from the film, *Planet of the Apes*, based on Pierre Boulle's book *La Planete des Singes*. The society depicted is one in which human culture has degenerated through laziness and their one-time servants, the apes, have taken over the rule of the land, refusing to grant them rights.

Ape Anatomy

Stripped of fur, the overall similarities between orangutans, gorillas, chimpanzees, bonobos, and humans are at once immediately apparent. Even with fur, we clearly have similar body shapes that differ greatly from those of other mammals, such as dogs, elephants, and whales. That said, while we and other apes are very similar, the differences between us also come into sharper focus when we are all naked, especially the relative proportions of the head, torso, arms, and legs, and the shapes and layouts of the nose, brows, chin, and other facial features. Delving under the skin, we can compare the muscles, skeletons, guts, brains, and other internal organs of wild apes with our own. The overall impression upon doing this is that they are all variations on a united theme. Anatomically, the apes are very much a close family.

Apes on the outside

We humans may be naked apes but some individuals have relatively short, sparse hairs, while in others the growth is quite luxuriant and thick. Just like humans, individuals in other ape species vary in the amount and distribution of their hair. However, the clearest difference in external appearance between them and us is that our 'fur' is much shorter and finer over most of our body.

Hairiness

A typical human body has over five million hairs, only 2 per cent of which are on the scalp. In most people, body hairs are short and fine, so much so that they are barely visible. Numbers of hairs among other apes are not all that different. It is the thicker, longer nature of their body hair that contrasts with our 'naked' appearance.

Oddly, humans outdo other apes in two respects concerning hair. Our 'fur' is longest on the scalp as head hair, and on the male's upper lip, cheeks and chin as the moustache-beard. No other ape's fur, on the head, body, or anywhere else, comes close to the length of uncut human scalp hair, which reaches more than a metre in many people and five times this in longest-hair record-holders. Only captive old male orangutans possess hairs approaching this length, as a hanging 'cloak' from the shoulders and arms. Such extravagant cloaks are hardly ever seen in the wild, since the hairs are trimmed or tugged out naturally when moving through the branches. When it comes to facial hair, only the male Sumatran orangutan has a beard that can claim to rival that of a male human. Again, this never attains the extraordinary lengths of some human beards, which may reach up to 5 metres (16 feet). Apart from head hair, why is our hair so reduced? Evolving in a hot open habitat (rather than among trees), needing to sweat after the physical effort of hunting, some type of aquatic phase in our prehistory, and minimizing homes for skin parasites have all been suggested as possible reasons, either separately or as a 'suite' of benefits.

Coloration

Human skin shows up well through our meagre fur. Its colour varies greatly from almost jet-black to extremely pale brownish-yellow or cream, as a result of adaptations to different temperatures, amounts of sunlight and other climatic conditions. This variation presumably developed as modern humans spread out of Africa and around the world within the past 150,000 years or so. Likewise our species' hair colour ranges from extremely deep brown-black to fair (and white in older individuals), all according to our genes.

The other apes are far more restricted in their distribution than us, with nowhere near the same range of climatic conditions, and as a result show far less variation in colour of their skin or fur. Even so, different populations have their subtle trademarks. These are more evident between species than within them. For example, the hair of

Sumatran orangutans tends to be longer, sparser, and paler than that of their Bornean cousins. In both species fur colour also varies with age, often being brighter orange or red when young, then darkening with age. Skin also tends to darken as individuals grow older, changing from pale pink or yellowish to darker brown, reddish, sooty-grey or even almost black in older adults, especially on the face and neck. Among gorillas, the darkest and most luxuriously furred is the mountain gorilla, with its thick black coat. Next is the eastern lowland gorilla. The western lowland gorilla is typically lighter and more variable in colour. The skin in both gorilla species is black or nearly so.

Chimpanzees and bonobos look broadly similar in terms of hair and skin. However, whereas bonobos tend to have black or very dark fur, chimpanzees show more variation, including shades of brown and steel blue. With the exception of the male human, all apes are largely lacking in facial fur. The ears, and the palms of the hands and soles of the feet, are also hairless.

On two legs
(Previous page) Standing seems an awkward pose to us for a gorilla, but then the human body's anatomy is heavily adapted for bipedal (two-legged) walking. We are just as awkward when going quadrupedal (on all fours), while the gorilla is fully at ease.

White ape
Albinism is exceptionally rare in all apes. It makes them conspicuous to predators, may affect their eyesight, and makes their skin more vulnerable to the sun's ultraviolet rays, which can lead to the skin cancer called malignant melanoma.

Red alert
The skin of the orangutan, known as the 'red ape' tends to darken with age, especially on the face, as shown in the contrast between this youngster and mother.

Size and shape

A large adult male gorilla can weigh more than 200kg (440lb). This makes him not only the largest ape but also the biggest of all living primates. At the other end of the great ape weight range is the bonobo. A petite adult female of this species can tip the scales at 25kg (55lb) or less.

Between these two ends of the spectrum are large areas of overlap in both weight and height, not just between ape species but also between subspecies, and within species and subspecies among both males and females. Overlain on all of this is weight variation in individuals with seasonal food abundance, as well as with age. As with humans, wild apes also vary in personality and their propensity to snack occasionally or gorge repeatedly when food is abundant.

Sex and size
Differences between male and female, known as sexual dimorphism, are well demonstrated by these western lowland gorillas. The male is not only larger but different in shape and proportions, for example, having a tall sagittal crest on the top of the head.

Bigger is – mostly – better

Among gorillas, the mountain gorilla tends to be the bulkiest, with wild adult males often weighing 160kg (350lb) or more. Eastern lowland gorillas are less hefty but taller than mountain gorillas. Western lowland gorillas tend to be shorter and lighter than their eastern cousins, with fully-grown adult males averaging 140–160kg (310–350lb).

Why are adult male gorillas so huge? Many theories have been advanced. One is that size is an advantage when defending the group against predators (chiefly leopards for gorillas) and when disputing territories with other gorilla groups. That said, in general, gorillas are not especially confrontational. They prefer to avoid their neighbours rather than clash with them. However if a face-off becomes unavoidable the males display their bulk and power and do their best to accentuate their size, as they attempt to intimidate the opposition.

Sexual selection may also play a part. Females tend to be most impressed for the above reasons by the largest males, who then father the females' offspring and so pass on their own genes for bulk. This process can lead to a great disparity in body size and shape between the sexes, known as sexual dimorphism. As females choose the bigger, more impressive partners they push up the average male size. Females themselves have their own selection pressures, such as being able to eat and process enough food for themselves and their developing offspring. In general, the bigger a female and her baby, the more food they need. In times of food shortage this can cause problems for large females, and this might explain why female gorillas have not followed the males by slowly becoming larger in size.

Although female gorillas are much smaller than males they are nevertheless large animals. Many scientists argue that they need to be the size they are in order to process their relatively energy-poor food. Gorillas are sometimes called 'primate ungulates' (ungulates are hoofed mammals such as rhinos, deer, camels, and cattle). Most ungulates need to be large because their bodies must obtain, digest, and process huge amounts of low-quality plant food, to furnish them with enough nourishment for survival. Gorillas have a similar diet of plentiful but poor-quality, difficult-to-digest herbage. Like ungulates, they spend much of their time foraging and digesting their food.

Gracile versus robust

Among orangutans, the Sumatran species is regarded as more 'gracile' – slimmer in build than the more 'robust' Bornean. There is little weight difference between the males or the females of the two species. Within each species, however, there is a large difference between males and females. The males have followed a similar evolutionary path to male gorillas. Big is best when impressing females and intimidating rival males. However, for the tree-dwelling orangutans, there is an upper weight limit decided by the strength of boughs and branches. Male orangutans are very close to this maximum, evidenced

Natural postures

The human ape's upright stance compared to the squat gorilla belies the latter's size and bulk. The difference between male and female within a species, or sexual dimorphism, is most marked for the great apes in the gorilla and orangutan. With a male human and orangutan each of 75–80 kilograms, the human female is 75–80% of this weight while the female orangutan is less than 50%.

ORANGUTAN	GORILLA	CHIMPANZEE	BONOBO	HUMAN
Male: (h) 130cm; (w) 78kg	Male: (h) 160cm; (w) 160kg	Male: (h) 125cm; (w) 55kg	Male: (h) 110cm; (w) 39kg	Male: (h) 175cm; (w) 78kg
Fem: (h) 110cm; (w) 37kg	Fem: (h) 140cm; (w) 80kg	Fem: (h) 110cm; (w) 40kg	Fem: (h) 95cm; (w) 31kg	Fem: (h) 162cm; (w) 59kg
Intermembral index: 140	Intermembral index: 115	Intermembral index: 104	Intermembral index: 103	Intermembral index: 80

by their wary progress through the canopy as they test branches for strength. Male gorillas have passed this point and gone from arboreal to terrestrial, spending much more of their time on the ground.

The commonly recognized subspecies of chimpanzee also vary slightly in overall size and build. The central chimpanzee is the largest and most robust subspecies, while the western and eastern chimps are both smaller and somewhat slimmer. Even more gracile and gangly are bonobos. This may be the result of neoteny – the retention of features from the young, developing body into full-grown maturity, so that adult bonobos in many ways resemble juvenile chimpanzees (see page 46).

Standing tall

The human skeleton and muscles are anatomically adapted for bipedalism. In other words, walking on two feet is our natural gait. The reasons why this came about are embedded in our prehistory (see page 66). The other great apes are capable of standing and walking on two feet, but it is not their natural posture. Their spinal column is at an oblique angle to the ground, not vertical like ours, nor horizontal as in most other quadrupeds.

For non-human apes, standing on two feet takes a great deal of muscular effort and their stability is compromised, especially in gorillas. When walking on two feet orangutans and chimpanzees, in particular, may hold out their arms to help with balance. The bonobo, with its longer legs and slightly more naturally upright stance, needs less effort to maintain this posture and spends more time standing and walking than the other hairy apes. In some surveys, bonobos have been found to spend as much as a quarter of the time they spend moving about doing so on two legs.

Bones, bodies, and limbs

The human body's skeleton of bones has been among the most iconic of all objects, depicted from stone-age cave art through the renaissance of anatomical portrayal to modern surrealism and ultra-detailed medical scans. Strip away the skin and flesh of our ape cousins and they have skeletons too – and the similarities are remarkable.

The overall pattern and arrangement of bones among great apes is almost identical, from the major bones of the skull, spinal column, and limbs to the three tiny ossicles or ear bones deep inside each ear. The human skeleton makes up about 15 per cent of our total body weight. In gorillas this proportion is 12–13 per cent; proportions of the other great apes vary between those for gorilla and human.

The human skeleton has 206 bones, and those of other apes are almost the same. One difference is that humans and orangutans have 12 pairs of ribs, generally speaking, while chimpanzeess and gorillas have 13. (There is natural variation in these figures: about 1 human in 13 has 13 pairs of ribs, for example, while a smaller proportion is born with 11 pairs.) The ribs are attached to a portion of the vertebral column known as the thoracic (chest) spine, with each rib pair joined to one thoracic vertebra (a vertebra is one of the several bones that make up the spine). Gorillas have 13 thoracic vertebrae and pairs of ribs, compared to our usual 12. The extra vertebra is in effect 'borrowed' from the portion of the spine below, the lumbar (lower back) vertebrae, so there are four of these bones in gorillas compared to five in humans. Although the usual figure is 206, human skeletons show individual variation in the numbers of bones overall, ranging from 200 to 210. The same applies to the other apes.

A myth persists that all men have one pair of ribs less (or more) than all women. This is entirely false. On average, the numbers of ribs and all other bones are the same for each sex, and this is also the case among the other apes.

Skulls

As befits the biggest apes, gorillas have the most massive skull. Like those of the other non-human apes it has projecting or 'prognathous' jaws and a massive mandibular ramus, the upright flange of the mandible (lower jawbone) that slots under the cheek bone and anchors the powerful masseter chewing muscles. The gorilla's skull also displays prominent eyebrow ridges and a ridge of bone along the top of the skull centre, from just behind the brows rearwards. This ridge is known as the sagittal crest and is another anchor point for the very strong temporalis chewing muscles that run down to the lower jaw, and are used for long periods as the gorilla masticates tough vegetation. In addition the gorilla skull has a nuchal crest – another ridge that runs around the upper rear of the skull, rather like a hair band that has slipped slightly backwards. This is yet another muscle attachment area, this time for the neck muscles. Orangutans also have projecting jaws, allowing them to process large mouthfuls of fruits. The main part of their lower jaw is especially deep and strong. The sagittal and nuchal crests are much less pronounced than in gorillas, as are the brow ridges.

The jaws are even less projecting in the chimpanzee and bonobo, with what could almost be considered a receding chin. The forehead is still sloping, but more defined than in gorillas or orangutans – the cranium (the upper 'braincase' part of the skull) is also a more significant structure beginning to dominate the upper head. There are still brow ridges, but again, these are not as manifest as gorillas'. They are even less obvious in the bonobo than the chimpanzee.

The human skull continues these trends with a high forehead and rounded cranium to house the enlarged brain. The face is almost flat, with no brow ridges and a reduced nose as far as bone is concerned (the flexible frame of the human nose is cartilage). The jaws are much reduced, hardly jutting at all, but there is an angular deep chin.

Torso

The gorilla's heavily built chest has thick ribs, with the lower pairs comparatively long in relation to the upper. On either side at the rear of the chest, the scapula (shoulder blade) is deep and sturdy, to support the shoulder joint and the very powerful shoulder and upper arm muscles. In fact, in most primates the scapula is more dorsal ('round the back' of the chest and nearer the vertebral column) than in other groups of mammals, where it is positioned more towards the side of the chest. This dorsal position allows much greater flexibility in the primate shoulder joint – a dog or cow, for example, cannot hold its front limbs out sideways, and could never lift them above its head.

Gorillas possess elongated dorsal processes (spine-like extensions of the vertebrae) in the neck region. These anchor the large muscles of the shoulder, neck, and upper back. The processes are less pronounced in orangutans, and even less so in chimpanzees and our own species.

At the base of the spine, the pelvis (hipbone) bears the hip joints and forms large flaring surfaces for attachment of the hip and thigh muscles. Non-human apes have an elongated, deep pelvis, which is narrower from side to side than the bowl-shaped human version.

Limb lengths

One marked difference among the apes is the proportional lengths of the front and rear limbs – the arms and legs. A useful comparative measurement here is the so-called intermembral index (IM), which is basically the ratio of fore-limb to hind-limb length. (More accurately, it is the added lengths of the humerus bone in the upper arm and the radius in the forearm, compared to the summed lengths of the femur in the thigh and the tibia in the shin.)

The human IM is about 80, indicating that the leg is one quarter as long again as the arm. This reflects how our legs have become longer and more powerful for upright walking, while our arms are no longer used in locomotion but have become adapted to carry and manipulate.

Chimpanzees and bonobos, by contrast, have IMs of around 100, showing that the arms and legs are approximately the same length (although the legs are slightly longer than the arms in the bonobo). The reduction in leg power is balanced by greater upper body strength for climbing. The bonobo's skeleton is generally slimmer and lighter-boned than the chimpanzee's, with a narrower rib cage, and limbs longer in proportion to the torso.

The IM of gorillas is 115, reflecting their greater arm reach and strength. The mountain gorilla has slightly shorter arms than the lowland gorillas, who have greater access to sturdy trees and hence climb more.

The most arboreal great apes are the orangutans, as suggested by their IM of 140. The arms are especially powerful and mobile at the shoulder joints, and the fingers almost touch the ground when walking on rear legs, even with the body relatively upright. By contrast our own fingertips are approximately level with the middle of the thigh.

Moving methods
As an orangutan stretches across the flooded swamp to a new tree, its long, strong arms show clearly in comparison to its shorter, less well-muscled legs.

Flexible friends
The chimpanzee's knees and hips are habitually flexed during knuckle-walking.

Skeleton and posture

The skeleton – whether of human, hairy ape, or other vertebrate – is the body's strong inner framework of bones, linked at moveable joints. The overall height, shape, and proportions of the body (excluding the effects of obesity) are determined by it. Among the most noticeable differences between the skeletons of the apes are skull size and shape, the relative lengths of the torso and limbs, and overall posture when standing on two legs.

HUMAN
Average skeletal height 170cm
Average body weight 70–80kg
Intermembral index 80

GORILLA
Average skeletal height 155cm
Average body weight 150–180kg
Intermembral index 115

Our own species has evolved to walk on two legs all the time, and this is reflected in subtle differences between our own skeleton and those of other apes. One of the most obvious differences to anatomists is the position of what is called the foramen magnum – the hole at the base of the skull through which the spinal cord (main nerve) passes from brain to body. In the non-human apes, especially orangutans, the foramen magnum is sited towards the rear of the skull, at an oblique angle so that the head hangs forwards. In humans the foramen magnum is almost below the centre of the skull and faces directly downwards. The human skull, with its heavy brain, therefore balances directly over the spine.

The hairy apes have a spinal column (backbone) which is gently bow-shaped when seen from the side, curving rearwards from the hips and then forwards again to the neck. By contrast, the human spinal column traces a gentle 'S' shape that allows the whole body – skull, chest, abdomen, shoulders, and arms – to balance neatly over the feet when we are standing up straight. This design reduces bone stress and muscle strain by transmitting body weight down the spinal column to its base, the sacrum – the wedge-shaped part that then transfers that weight to the pelvis.

The human pelvis is more bowl-shaped than that of other apes. This is another adaptation for standing and walking upright: our body weight is transferred through the sacro-iliac joints (the joints between the sacrum and the left and right parts of the hipbone) to the legs. The other apes have a longer and narrower pelvis. Viewed from the front, the femur (thighbone) of the non-human apes is almost vertical, so that a standing ape's legs are more or less parallel with a gap

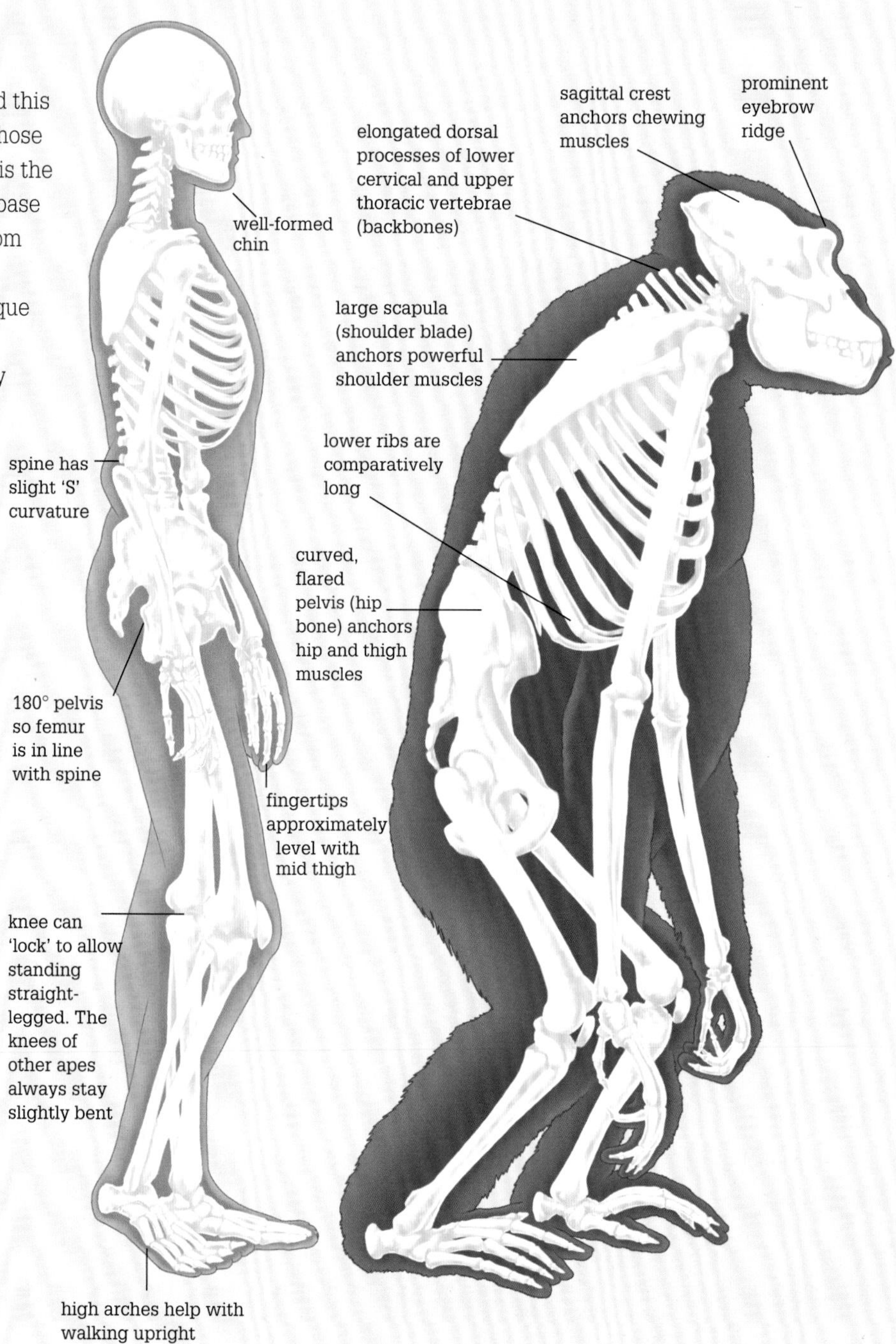

Skeletons compared

Side views of male great ape skeletons demonstrate the human's balanced upright posture, head and hips and feet all in a vertical 'tower'. In the other species the neck part of the backbone joins the skull at an oblique angle from the rear rather than from directly below.

between the knees. The human femur is different – it angles inwards from hip to knee, due to the structure of its 'head' and 'neck' at the upper end, its curving shaft, and the knee's anatomy at the lower end. This allows a human to stand upright with knees together and knee joints 'locked'. Our anklebones, of which there are seven on each side, are bigger and stronger than those of the other apes, because of their weight-bearing role. Our feet have many structural modifications for walking, at the expense of the flexibility and dexterity seen in the feet of other apes. These skeletal differences are not the only ones between our species and other apes as a result of our bipedalism. There are also various differences in the joints, ligaments, and muscles.

Stories bones tell

The detailed structures of bones from different apes are almost indistinguishable. Most ape bones have an outer shell-like layer of hard bone tissue, where the main strength resides. Under this is an intermediate layer that is more sponge-like or honeycombed, while in the middle is the marrow cavity, a central chamber filled with jelly-like bone marrow. The marrow is biologically extremely active, storing fats and other nutrients, and manufacturing new blood cells by the million every minute, as part of the blood's natural cellular turnover. So the skeleton is not just a jointed inner framework – it is a store of valuable minerals, and a busy 'blood factory'. Skeletons of dead apes provide a great deal of information about their owners' life experiences, from periods of hunger to broken bones, infection, and disease. Following examination of their skeletons, the chimpanzees of Gombe, Tanzania, have been found to be smaller than those from many other sites, for example. Because Gombe is a drier area than, say, Taï in Côte d'Ivoire, another studied site, this may be linked to the Gombe chimpanzees' inferior diet, lower in nutrients than that of most other chimpanzees. Chimpanzee and other ape skeletons also reveal many instances of cracked and broken bones that have subsequently healed. Flo, the famous matriarch from Gombe, who died in 1972, was found to have suffered fractures to her clavicle (collarbone), arm, and several toes, and fingers. Injuries like these usually result from falling out of trees, but some are known to be the result of fighting with other individuals.

ORANGUTAN
Average skeletal height 120cm
Average body weight 70–80kg
Intermembral index 140

CHIMPANZEE
Average skeletal height 120cm
Average body weight 50–60kg
Intermembral index 104

BONOBO
Average skeletal height 105cm
Average body weight 30–40kg
Intermembral index 103

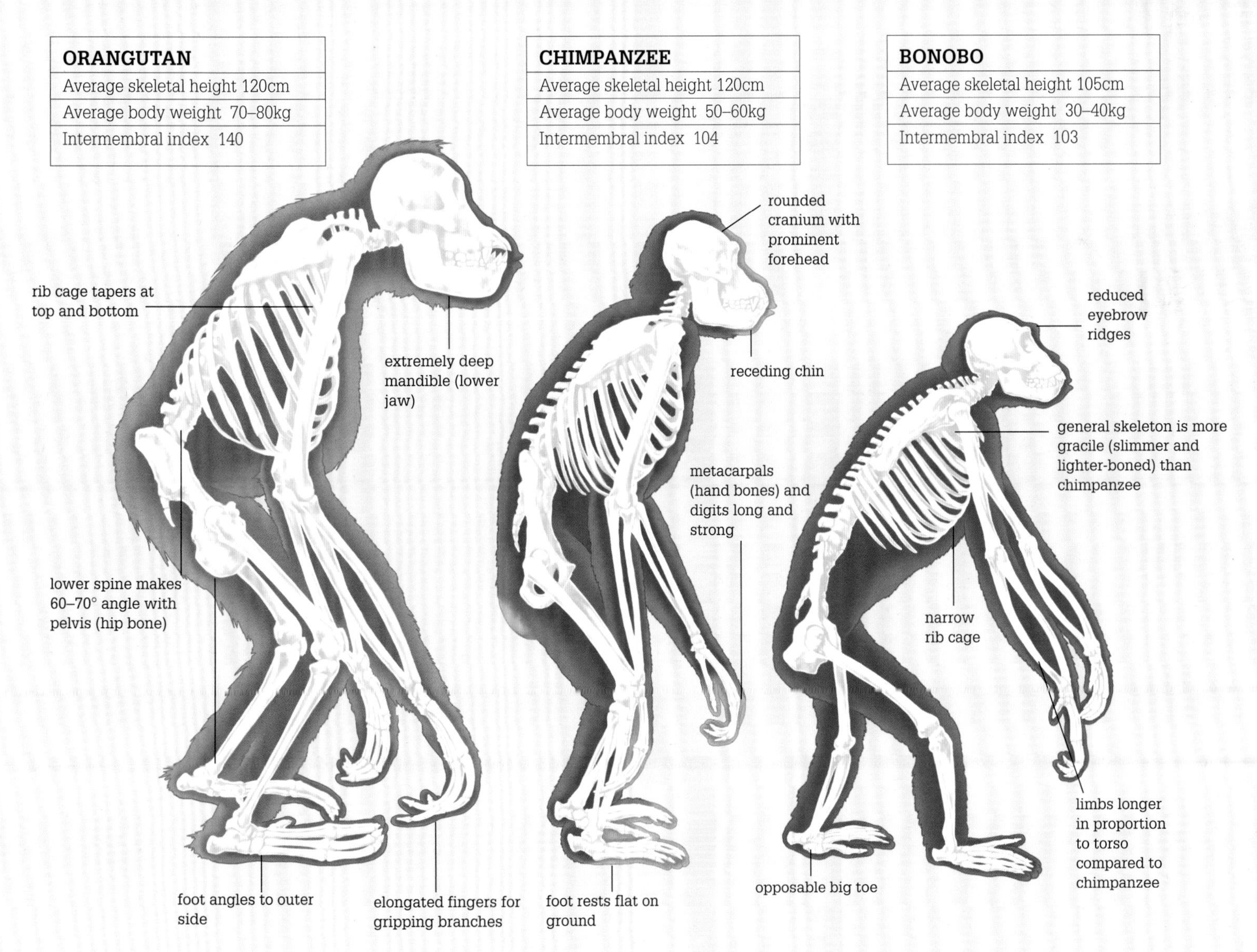

Muscles and moving

Stories abound of how muscular apes outdo puny humans in trials of strength. But direct comparisons of power and strength between the ape species are exceptionally difficult to arrange.

Many of the often-quoted 'assertions' are based on a study undertaken in 1924 at Bronx Zoo, New York, using a spring-based mechanical contraption called a dynamometer that measured force of pull, in a similar way to weighing scales that measure gravity's downwards force. A human male volunteer weighing in at 75kg (165lb) achieved a one-armed pull of 91kg (200lb), with several other males averaging 80kg (175lb). A male chimp named Boma of the same body weight put this in the shade by registering 385kg (850lb) one-handed. A female chimp, Suzette, then achieved an amazing 572kg (1,260lb), although she used two hands and was very agitated at the time. The human volunteers averaged 170kg (375lb) two-handed.

From this study and others like it grew the belief that size for size, a chimpanzee is four to seven times stronger than a human, an orangutan five to eight times stronger, and a gorilla as much as twelve times stronger. While these figures are not strictly scientifically proven there is plenty of evidence, both anecdotal and formal, that the other apes are more powerful than we are, at least in terms of their arms and upper body strength. The reason for this is obvious when one considers their lifestyles. Wild apes use their arms much more for locomotion and heavy-duty moving than we do, so their muscle mass, bone sturdiness, and upper body strength need to be much greater.

Muscle structure

Muscles are basically bundles of hair-thin, thread-like units called muscle fibres or myofibres. Each myofibre consists of even smaller, thinner threads, known as myofibrils. Studies of the jaw muscles of our own species and those of other apes, as well as those of various other primates, reveal that human jaw muscle fibres are smaller and weaker. The reasons for this are not clear. However, genetic research has turned up one fascinating nugget of information. The genes that carry instructions for the ongoing construction of myosin (to keep muscles well built and in good working order) are identical in most primates, including various monkeys and all the non-human apes – but not in humans. We have a modified or mutant version of one gene (known to genetic scientists as MYH16) that instructs for a weaker, less active type of myosin. So the molecular structure of our muscles is slightly different. As a result, human muscles are simply less powerful than those of most other primates and all the other apes. This discovery has led to some interesting speculation. Powerful muscles need strong

Major muscles of the shoulder, lower back, and hip region compared in human and chimpanzee

Trapezius runs from upper spine to scapula (shoulder blade), to steady the shoulder; large in the chimpanzee it reinforces arm movements especially in upper part to help arm-raising.

Latissimus and erector help to steady and support lower back, especially when flexing the spine forward and to the side; large in human for upright stability.

Gluteus maximus is the main buttock muscle, very enlarged in humans compared to other apes. It provides the main force to pull the thigh bone back at the hip to thrust the body forwards when walking, running, leaping.

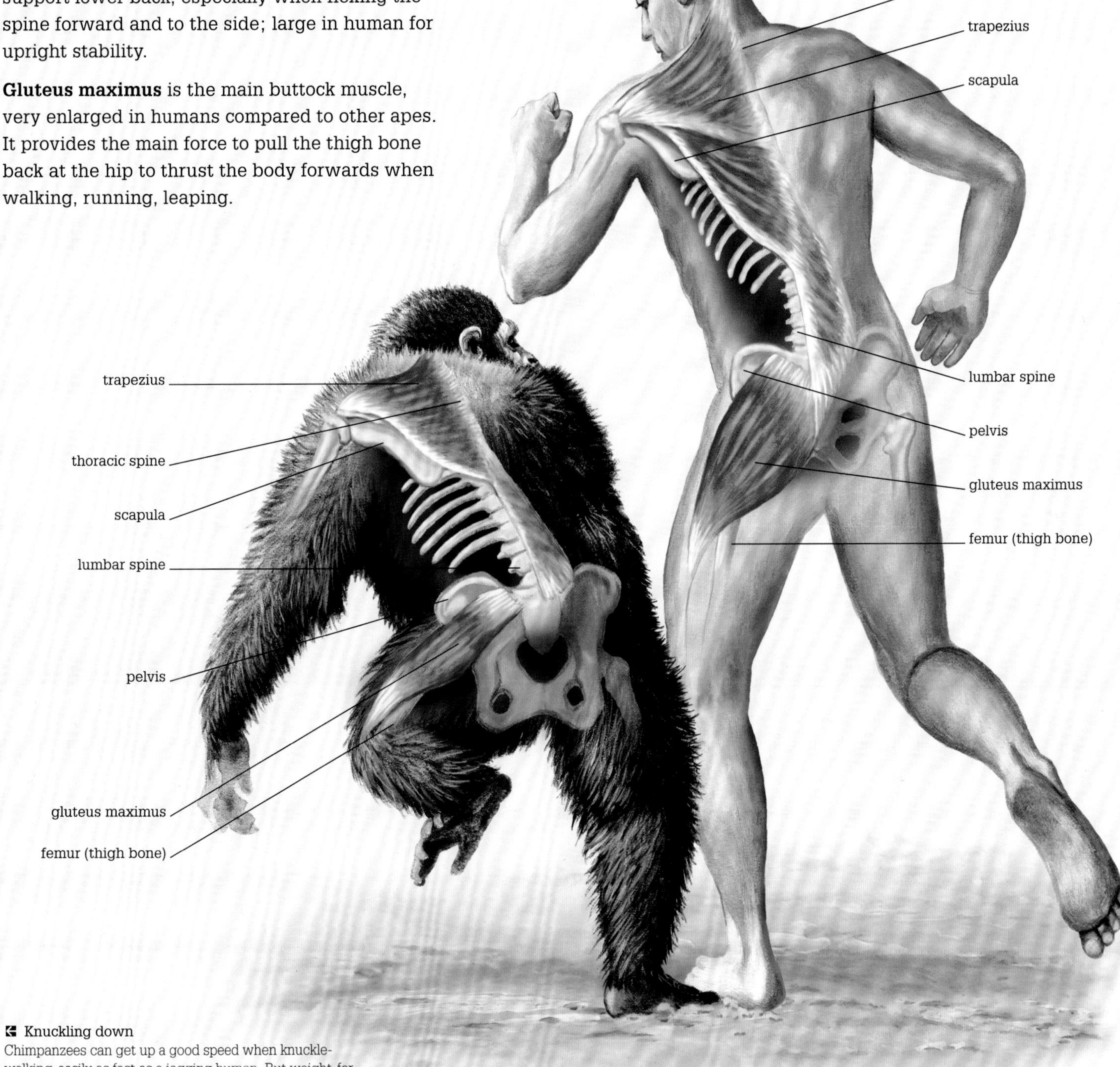

Knuckling down
Chimpanzees can get up a good speed when knuckle-walking, easily as fast as a jogging human. But weight-for-weight, for the distance covered, they use about four times as much energy as an upright-walking human.

bones to anchor them and pull against. After the altered human myosin gene cropped up, possibly millions of years ago, weaker muscles perhaps released the bones around them from the arduous task of staying strong and sturdy. In the skull, this would have allowed the bones a certain degree of freedom to change and expand – for example, to accommodate an enlarging brain. This line of thought ties in with the loss of the sagittal crest and other ridges on the human skull. It is possible that our braininess developed as a side effect of a mutant muscle gene. Whether or not this was the case remains to be proven, but the idea is the subject of much genetic and biochemical work.

On the ground

Non-human apes are basically quadrupeds. They move around using all four limbs. Gorillas, chimpanzees, and bonobos indulge in knuckle-walking, where the fingers are curled over in a partial fist to take the forward portion of the body's weight. The force is not on the knuckles proper, but on the joint between the end and middle bones of the fingers, and to a lesser extent between the middle and inner bones. As a consequence, it is the intermediate phalanges that bear much of an ape's weight when it is on all fours. The second joints of the fingers are covered with thick skin to cope with the increased pressure and wear.

When moving on all fours a gorilla's back forms an oblique angle to the ground, mainly because its arms are much longer than its legs. The back of a chimpanzee knuckle-walking forms a less oblique angle and a bonobo's may be almost horizontal. Orangutans, with their proportionally longer hook-like hands, sometimes indulge in 'fist-walking', where the fingers curl over further and weight is taken on the finger bones next to the palm. They also palm-walk, holding the hand palm-down against the ground – a bit like a crawling human baby, except that the fingers are curled up rather than pointing out straight.

Energy expenditure

How do the other great apes measure up when they walk upright? Experiments with chimpanzees trained to use a treadmill showed that they expended much more energy for their size than humans did. Given that bipedal walking is not their normal mode of locomotion this result is what one would expect. What is surprising is that chimpanzees use roughly the same amount of energy whether they are moving bipedally or quadrupedally, and that, weight for weight, it is four times the amount that a human uses. Human bipedal walking, it would appear, is very energy efficient. As we stride out to cover long distances, two legs and an upright, balanced gait are superior when it comes to energy conservation.

Humans are competent high jumpers, but what about the other great apes? A recent project compared captive bonobos with human athletes in their ability to jump vertically from a still standing position. Perhaps surprisingly, the bonobos won hands down, or rather, feet up. They could spring directly up almost 80cm (30 inches), nearly twice as high as the humans. Even taking into account the weight difference – 34kg (75lb) for a male bonobo and 61.5kg (135lb) for his human opponent – the bonobo's muscles generated far more mechanical force. And all this from a slender, gracile, 'pygmy' cousin of the chimpanzee!

In the trees

Orangutans are in effect quadrupedal climbers, using their hook-like hands and feet to grasp branches as they travel through the rainforest canopy. Their shoulders and in particular their hip joints are especially flexible, with looser ligaments in the hip holding the head of the femur against the pelvis than are found in other great apes. On top of this, the way that the forearm muscles and ligaments are arranged produces an amazing range of wrist movements. Orangutans also travel bipedally in the canopy, walking on their feet while extending their arms for balance and to grasp passing boughs. Generally speaking, the wider a branch is, the more likely an orangutan is to walk on all fours along it. This suggests that, on narrow objects, orangutans have better balance when standing upright than they do on all fours.

Whether on the ground or in a tree, an orangutan can manipulate both its hips and knees until they are relatively straight, to produce a more upright posture than one might expect. Bonobos also have this ability. In contrast, chimpanzees tend to keep both their hips and knees bent and their torso angled forwards. This results in a bent over, shuffling gait.

Recent observations of orangutans have shown that they move more often on all fours than they do bipedally, with the body held upright. This is true whether they are in the branches or on the ground. Another finding is that 'suspensory locomotion', where the body weight hangs from the limbs, is less common in orangutans than either quadrupedal or bipedal walking. It seems that the traditional view of an orangutan swinging along by hanging from its arms – a form of locomotion known as brachiation – is much less common than we might imagine. The masters of this form of movement are the gibbons, smaller apes that in some places share the orangutans' habitat.

Catch me if you can
Young mountain gorillas play chase and tag as they learn the ropes of climbing.

Learning to climb
Infant orangutan orphans have to learn to climb without the constant example of the mother providing a tutorial masterclass. A one-handed swing may seem fun, but it's a risky manoeuvre at only two years of age.

Faces

Facial features show even more variation than fingerprints, giving us the information we need to recognize individuals, and they also contribute hugely to the way that we and the other great apes communicate visually by expressions, as explained in more detail on page 166.

A cursory glance at a group of apes from the same species, especially if they are on the move, may well elicit the reaction from a human that 'they all look the same'. But a gallery of photographs of ape faces immediately shows this to be false: they are as individual in terms of appearance as we are. That said, the other apes do not have the range of facial skin colours seen in humans from around the world. Their skin colours are more conservative and predictable for each species. But the relative size and proportions of the forehead, eyes, nose, cheeks, lips, and 'chin' are endlessly varied among the individuals within the various ape species. As in humans, even wild ape identical twins (which are rare, but not unknown) grow up to look slightly different from one another. As with ourselves, facial appearance and the relative size of the head to the body is linked to being male or female. It is also linked to age, from the proportionally big-headed infant to the larger-bodied and proportionally smaller-headed adult or old-timer.

Bonobo or chimpanzee?

Bonobos are often thought of as slightly slimmer but otherwise not much different to chimpanzees. A closer look reveals many more differences. Bonobos are born with dark facial skin, although the skin's shininess and ambient lighting may make it appear paler than it is. Baby chimpanzees have pink skin that darkens with age. A bonobo's lips are usually light red or pink; a chimpanzee's brown or black.

The bonobo's skull is more rounded than a chimpanzee's, shorter from front to back, and more prominent in the forehead area. It also lacks the large brow ridges, which give chimpanzees a more serious look. The chimpanzee's skull is longer with more pronounced brow ridges and a lower forehead. Looked at face-on, a bonobo's nostrils are wider, and its eyes are set farther apart and less deep in the face than a chimpanzee's. The typical adult chimpanzee develops something of a white beard on the chin as it ages but rarely grows white side whiskers. In the bonobo the reverse is true. Another difference is the bonobo has a natural centre parting in the hair on its head. This can appear so neat that it looks as if it has been carefully brushed in the mirror just a few moments previously. Within the chimpanzee species, facial appearances differ among individuals but there are facial trends according to subspecies. The western chimpanzee has a shorter, thicker, more rounded beard than the eastern chimpanzee, which grows a more straggly effort. Also the western chimpanzee's head is less broad, with more accentuated brow ridges and a steeper, more vertical occiput (the rear-facing portion of the head).

Darkest and palest

Gorillas are the most heavily pigmented of all the apes, with normally jet-black skin in the facial and other hairless areas of the body. Like the other non-human apes, their eyes lack the range of blues, greens, greys, and other colours seen in our own species, and instead tend to shades of brown. Eyes and nostrils are set wide apart in the massive skull, the muzzle is short, and the forehead recedes rapidly up from the beetling brow ridges, sometimes to a point-like crown. The brooding, powerful appearance of the face is offset by the small deep-set eyes and relatively small ears. Mountain gorillas tend to look slightly less intimidating than their lowland cousins because their longer, thicker fur gives their heads a more rounded appearance and makes the areas of hairless skin on their faces appear smaller than they actually are.

The ape with the greatest growth of facial hair is the human male, although some orangutans come close (see page 100). The Bornean orangutan generally has less fur around the face, which is rounder in outline and more dish-like in shape than that of the Sumatran orangutan. The Bornean orangutan's face also has a more hourglass or figure-of-8 appearance, due to its suborbital fossa – a depression or pit-like area in the skull below the eyes.

The Sumatran orangutan's face is longer and has longer, paler facial hair; the females may have a pale or white beard, and most males develop a prominent beard and moustache. The suborbital fossa is less noticeable, making the sides of the face appear straighter.

In both species of orangutan males develop fleshy cheek pads, called flanges or flaps, as a sign of sexual maturity. They make their owner look more intimidating to competing males, and may also help to amplify long-distance courting calls. That said, not all males do this. A situation known as bimaturism sees some sexually mature males in a sort of 'arrested development': they fail to develop the outward 'secondary' sexual characteristics for up to 20 years. Recent observations have shown that both the fully-flanged males and the smaller unflanged ones can reproduce, but that they use different strategies (see page 219).

Orangutans have brown eyes, like the other non-human apes. However, the white of the eye, around the coloured iris, tends to be visible more often in orangutans than other non-human apes. Its visibility can give the impression of a more 'human' gaze.

Going somewhere important
(Previous page) An adult female western lowland gorilla strides out purposefully, knuckle-walking as her youngster clings on in typical piggy-back fashion. Her huge shoulders and arms have almost twice the muscle mass of the hips and legs.

Studying faces
Compare each facial feature in turn in these four great apes. Opposite and below, left to right: The Bornean orangutan's nose is small with close-set nostrils; the mountain gorilla's is large with wider-set nostrils; the chimpanzee's nose is fairly flat and reduced; the human nose is by far the most protuberant, with downward-facing nostrils.

Hands and feet

All apes have much the same basic bones in their hands and feet. There are five metacarpals in the palm of the hand, and five equivalent metatarsals in the sole of the foot. The first digit has two phalanges (finger or toe bones), proximal and distal, forming the thumb (known anatomically as the pollex), and big toe (hallux). Orangutans differ in usually having just one toe bone. The second to fifth fingers and toes in all apes have three bones, known as the proximal, medial, and distal phalanges – the distil phalange is the outer one, nearest the tip.

Like most of the primates, including ourselves, the great apes have nails at the tips of their fingers and toes, instead of the claws or hooves found among the vast majority of other mammals. Nails are not just for scratching and picking off fleas. They provide a firm 'backboard' for the fingertips, allowing pressure on the skin of the underside of the finger to be gauged more accurately by the brain. This is important in judging the strength of grip, allowing apes to apply types of grip to branches, food, or the hairs of a grooming partner. Nails also strengthen and protect the fingertips from damage due to physical impact.

All apes have mostly bare skin on their palms and the palm sides of their fingers, as well as on their soles and the sole surfaces of their toes. The skin in these places is ridged for touch sensitivity and for grip, which is especially important when moving through slippery branches. These tiny ridges form uniquely individual fingerprints and toe prints, both in our own species and in other great apes. While human fingerprints can be used to detect criminals, those of other apes also have their uses. Orphaned baby orangutans are routinely fingerprinted (and microchipped) at rescue centres, for example.

Manual dexterity

As well as the above similarities, there are also plenty of differences between the hands of apes. These are most pronounced between the hands of our own species and those of our hairy cousins. We have much longer palm bones or metacarpals than other apes. Our palms and fingers are straighter, and the thumb in particular is proportionally longer. As a result we are by far the most dexterous and manipulative apes, able to move each finger with considerable flexibility and the thumb even more so. We can also touch each fingertip with the thumb, an ability known as opposition. All this is the result of our hands being freed from the chores of bodily movement or locomotion. We no longer have to knuckle-walk or climb, so our hands are less hook-like and the skin on them is much thinner, allowing for far more subtle movement.

Among the other great apes, chimpanzees show greatest manual dexterity. They use their hands for a wide range of tasks beyond simply gripping branches and holding food.

Multipurpose hands
Holding the crooked fingers together forms a shallow bowl, enough for this orangutan to scoop up water for a refreshing drink – and show her infant how it's done.

Navel gazing
A chimpanzee delicately picks at skin and hair as part of the grooming process, important both hygienically and socially.

Hands and feet compared

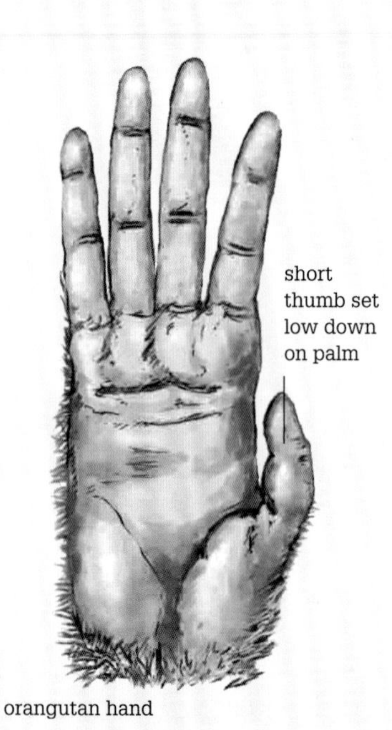

orangutan hand

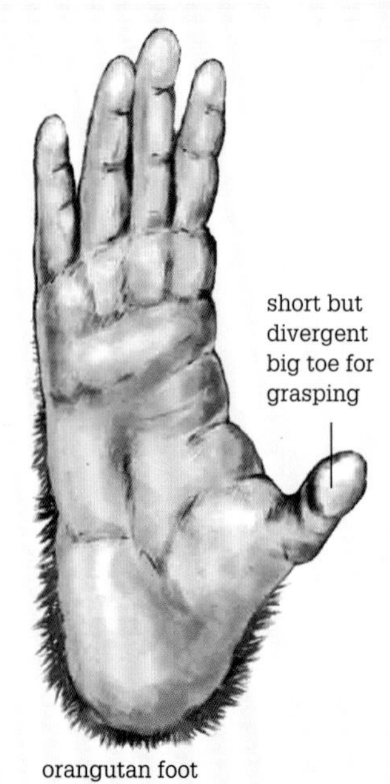

orangutan foot

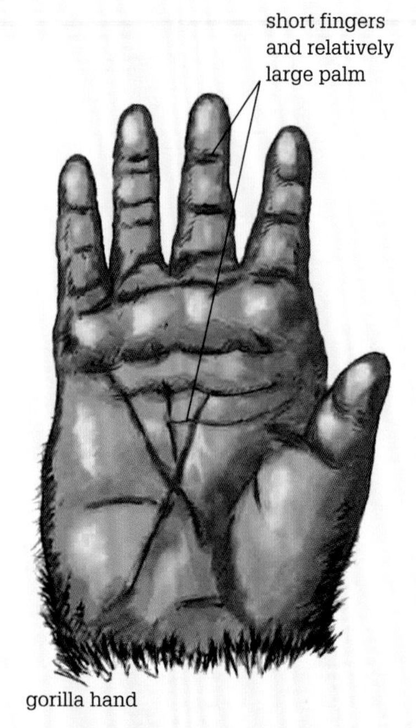

gorilla hand

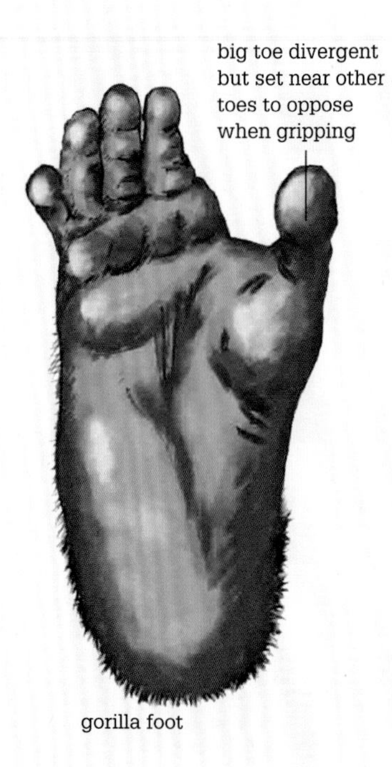

gorilla foot

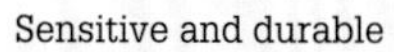

Gripping the round
The orangutan's hand ↓ and foot ↘ show how the very long fingers curl right around creepers, vines, and branches for a secure grip.

Sensitive and durable
A gorilla hand ↓ can be surprisingly dextrous, despite the fingers' reduced length. On the foot ↘ the robust, mobile big toe opposes the other four stubby but stout toes, and the skin is thick and tough.

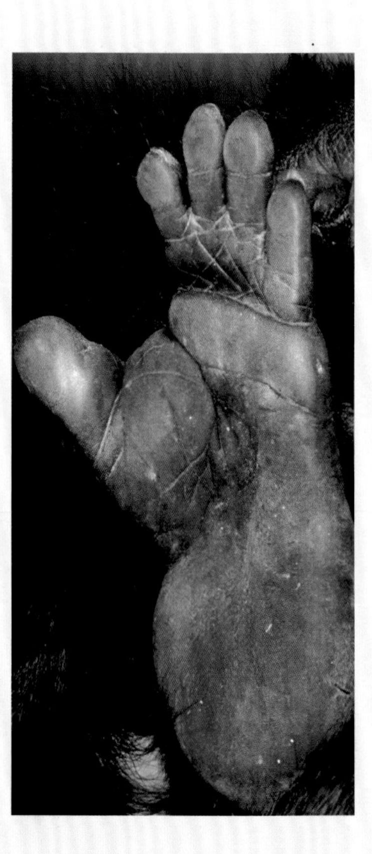

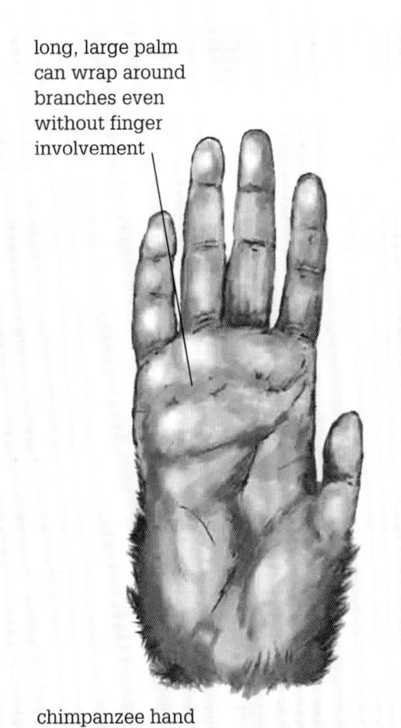

chimpanzee hand

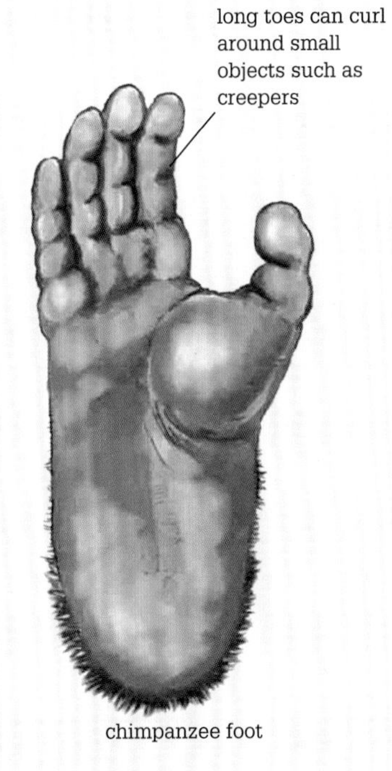

chimpanzee foot

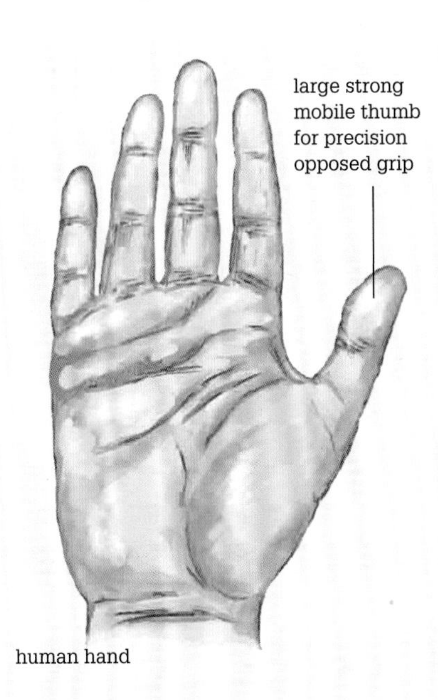

human hand

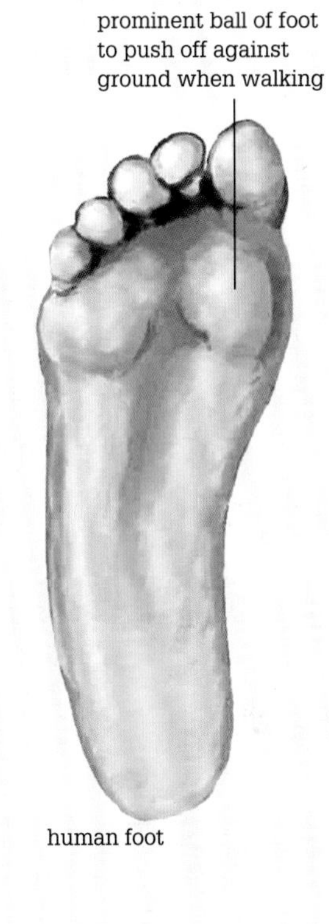

human foot

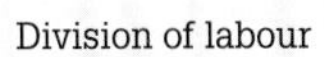

Power grip
The chimpanzee's hand and foot are equally adept at powerfully grasping, with the thumb opposing the fingers into the palm or the big toe opposing the other toes into the sole.

Division of labour
Compared to the other great apes, the human hand and foot have diverged greatly in anatomy and mobility. The hand manipulates accurately with its thumb-forefinger precision grip, while the foot (below, right) is adapted for weight-bearing and balance adjustment.

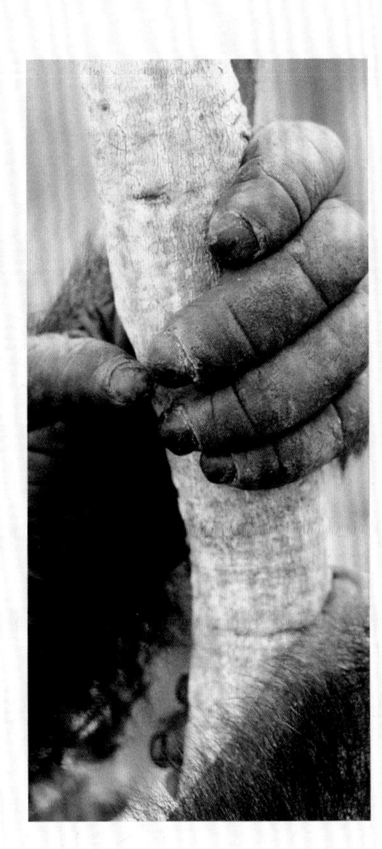

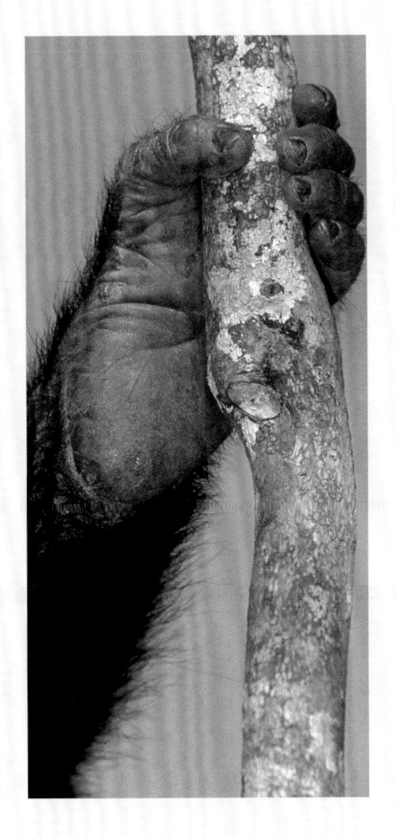

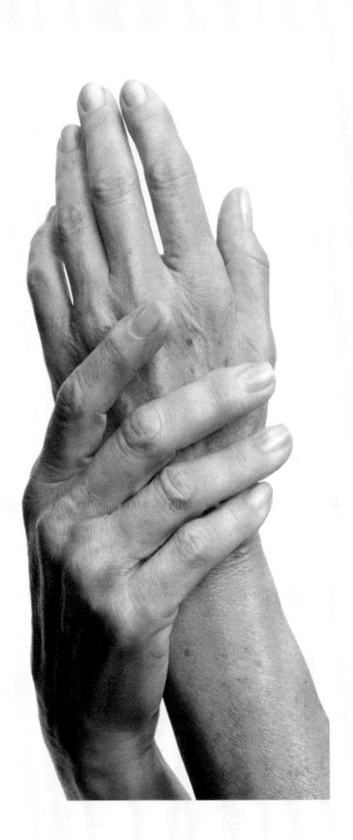

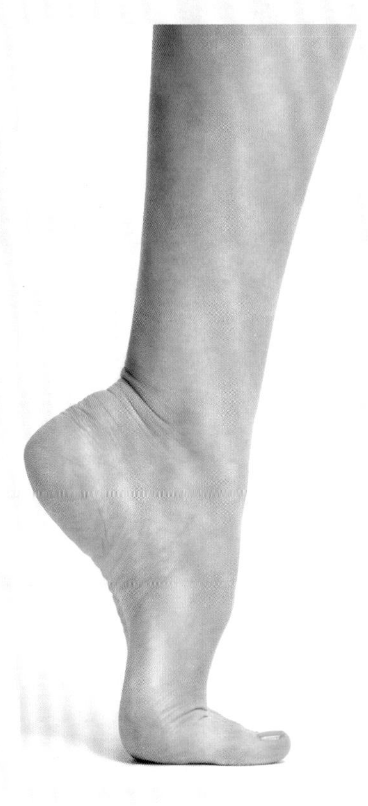

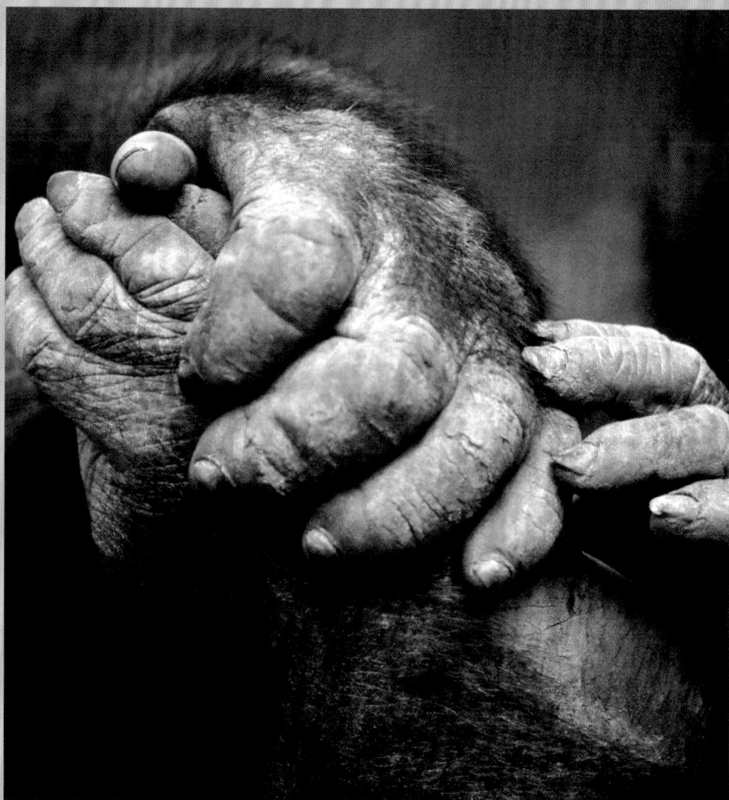

Manipulated objects range from brandished sticks and other weapons to delicate tools for feeding. The hands are also used for removing gravel, sand, or mud from food and tiny pests from the skin and hair when grooming (self or mutually).

However, a chimp's hand is nowhere near as dexterous as our own. The chimpanzee's hand musculature is greater and its joint flexibility less. On top of this, the relatively small size of the thumb and larger palm mean that its thumb is less opposable than ours. Its tip cannot precisely touch the tips of the other four fingers in turn. Importantly, this means that chimpanzees cannot generate the same squeezing power between finger and thumb as us. This means that chimpanzees cannot hold very small objects as tightly or delicately as we can.

Getting a grip

Like our own hands, a chimpanzee's hand has two basic grasps: a precision grip, used when holding a small object such as a pea, and a power grip, used when grabbing a larger object such as a rod or branch. The precision grip allows an object to be rotated and angled by movements of the fingers, using the thumb as an opposing backstop. Compared with our own precision grip, the chimpanzee's precision grip is greatly limited by the small size and relative weakness of its thumb. This is set low on the palm, due to the greatly reduced size of the first metacarpal bone (which is within the palm itself). A chimpanzee's precision grip usually involves the thumb holding the object against the side of the second digit or forefinger, rather than against its inner surface or its tip, as we would. The other fingers have minimal involvement in the precision grip.

The power grip, where the held object nestles between the base of the thumb and the second digit (the forefinger), imparts much more strength to the grasp. This grip is the basis of most of a chimpanzee's handling actions. When holding an object like this the other fingers are usually curled around it. While this gives the grip much more strength, accurate manoeuvring of the object is greatly limited.

A chimpanzee's thumb plays a lesser role in the power grip than our own does. For most jobs chimps tend to use a hook grip, with four bent fingers and no thumb involvement at all, such as when hanging from a horizontal support. The second and third digits (the forefinger and middle finger) are held in a natural curve when not in use. This is a result of their frequent involvement in hook and power gripping.

Compared to those of other apes, gorillas' hands are much shorter and wider, relative to their body size. This probably reflects their mainly ground-living rather than arboreal lifestyle, with less need for the hook grip, used for holding onto branches.

Flexible hooks

The muscles that flex the fingers, curling them towards the palm, and those that extend or straighten them are in the forearm, connected to the fingers by long tendons running through the wrist. An orangutan's finger-flexing muscles are four times the bulk of its finger extensors. In the African apes the finger-flexing muscles are comparatively less bulky. This reflects the power of the orangutans' hook grip, a result of their almost totally tree-dwelling existence. Orangutans also have very flexible wrists, capable of swivelling through 150° (the African apes can achieve 110–120°). The very short thumb – less than one-third the

total length of the hand – has weak thumb-adducting muscles that curl it towards the fingers. As a result, orangutans have a relatively weak power grip. However, their long, flexible fingers confer a different advantage – a 'double-lock' grip. Here the finger curls around so much that its last bone, the terminal phalanx, comes to lie parallel to the bones of the palm, allowing the fingernail to press flat against the palm skin. In this position the finger is 'locked' against the palm. Using this grip, objects with a relatively small diameter, such as thin branches, can be gripped extremely tightly with very little muscle effort.

Hands or feet?
A chimpanzee holds its foot (in the centre) with its hand (coming in from the right). The toes are similar to the thumb.

Gorilla grip
The gorilla's fingers are stubby and powerfully muscled.

King of the swingers
The orangutan's long fingers curl easily into a hook shape for branch-holding.

Fingers and toes

The anatomical differences between the human hand and foot mean we could never mistake the two. Our big toe is aligned with the other toes and has very limited powers of movement. Unlike the big toes of other apes it has no opposability. Our ankle bones are bigger and stronger than those of other apes in order to support our weight when walking. However, this strength comes at a cost – they are much less mobile. Our springy foot arch, with its energy-storing ligaments and tendons, is completely lacking in the foot of other apes. We also have a long sole and relatively short toes. Again, this is an adaptation for walking rather than climbing or gripping objects.

Among other apes, and especially in orangutans, the hand and foot are much more similar to one another. Orangutan toes are able to curl and grip extremely well, and the big toe is divergent (set at an angle away) from the other toes so that it can assist in a variety of gripping movements. An orangutan's foot is so long that it makes up almost one-third of the whole hind-limb length, and its toes are four-fifths the length of its fingers. Both of these proportions are more extreme than in the African apes and reflect the orangutans' largely arboreal lifestyle. Unusually, the big toe of orangutans normally consists of just one bone or phalanx, rather than the two seen in other great ape species. As a result, the 'big' toe is in fact the smallest of the five toes on an orangutan's foot and is set so far down the sole that it might almost go unnoticed. Nevertheless, the orangutan big toe plays an important role in climbing. Set almost at a right angle to the other toes, it can curl around in the manner of our own fingers, allowing the whole foot to grip extremely well. On top of this, an orangutan's ankles, like its wrists, are highly mobile, allowing the foot to angle inwards or outwards. The individual toes can both be lifted up or curled down towards the sole. The consequence of all this flexibility is that an orangutan might be considered 'four-handed', able, as it is, to support its entire body weight hanging from either a single hand or foot.

Brains and nerves

In one respect, much of this book is about the ape brain – the seat of mental processes, memories, behaviour, and what we might term emotions, intelligence, and understanding. But in terms of gross anatomy, or the brain structure as seen with the naked eye, how do the great apes measure up against each other? And how much do we know about how their brains function?

The main parts of the brain are similar in all apes. Well protected within the skull, this organ is dominated by the cerebrum – the immediately obvious wrinkle-surfaced upper part. The cerebrum is pinky-grey in colour and has a consistency not unlike blancmange. It is divided by a deep groove into left and right halves, known as the cerebral hemispheres. Their outer surface, the cerebral cortex, or grey matter, is the main site of higher mental processes such as thought, planning and executing actions, learning and memory, and of the tendencies, likes, dislikes, and other qualities that form the 'personality' – a term which can be applied to all apes, not just humans. The cerebrum wraps around the 'core' of the brain, or brain stem, whose lowest part can be seen projecting from the bottom of the organ. The brain stem's upper areas are involved in lower-level mental processes such as awareness, biorhythms (the daily and other time-controlled cycles of the body), sleep and wakefulness, and the daily cycle of behaviour. It is also the seat of basic drives such as hunger, thirst, fear, rage, and sexual motivation. The brain's lower region, the medulla, harbours the control and coordination centres for life-support processes including heartbeat, breathing, digestion, and excretion.

At the lower rear of the cerebrum, behind the brain stem, is a smaller wrinkled part, known as the cerebellum. This is heavily involved in physical movements, and especially in coordinating muscles so that well-practised actions become smooth and skilful.

Nerves carry sensory information from the eyes, ears, nose, tongue, skin, and balance organs to the brain, and also convey motor information from the brain out to the muscles, to control their contractions and movements. The brain is linked to the eyes, nose, ears, and other parts of the face, head, and neck by pairs of cranial nerves that join onto its underside. The lowest part of the brain stem narrows to become the body's major nerve, the spinal cord. This passes along the inside the backbone or spinal column through a row of holes in the vertebral bones. At the joints between these bones, spinal nerves branch out from the spinal cord to all of the body parts.

Is bigger better?

Are bigger brains more intelligent? Not necessarily. If they were, then the sperm whale, with its 8kg (17.5lb) brain (more than five times heavier than ours) would be the smartest creature on please . Brain-to-body mass ratio (also known as the Encephalization Quotient or EQ) is in some ways a better indicator of intelligence. Humans measure up at about 1:50 (that is to say our brains make up about one-fiftieth of our body weight), greater than any of the other apes. The small monkeys known as marmosets come in at 1:40 or less, with baboons, squirrel monkeys and mangabeys even lower. A third criterion which must be considered when trying to assess relative intelligence is the overall surface area of the 'grey matter' or cerebral cortex, and the interconnectedness of its billions of nerve cells. The human brain comes out top among primates in both of these measurements.

Idle thoughts

Scanning techniques from modern medicine such as PET (Positron Emission Tomography) and fMRI (functional Magnetic Resonance

Scent sense
The sense organs detect features such as light rays, sound waves and odour molecules, and generate corresponding patterns of nerve signals for the brain. It is in the brain that we become aware of the sensations.

Brain to body ratio

Humans stand out as anatomically the 'brainiest' of great apes, both in absolute brain size and in the ratio of brain to body. For comparison the elephant is 1:1,000, the dolphin about 1:100. Some smaller monkeys like marmosets are 1:30 or less.

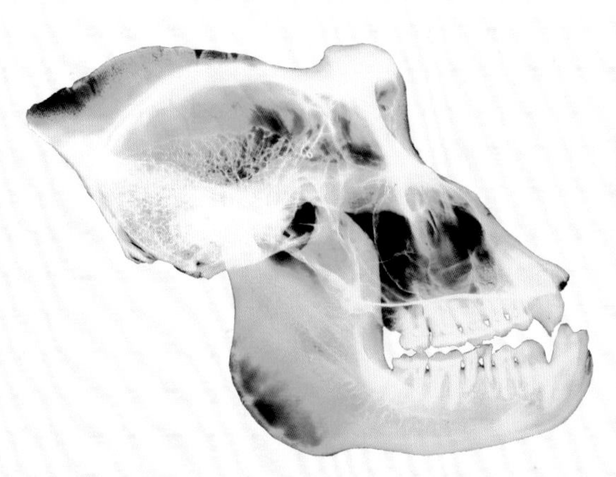

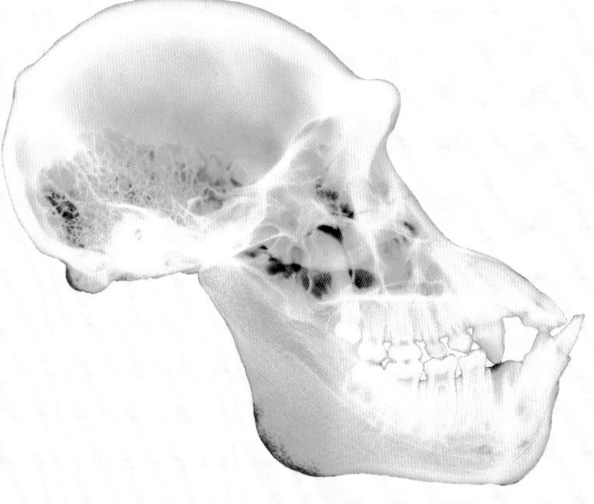

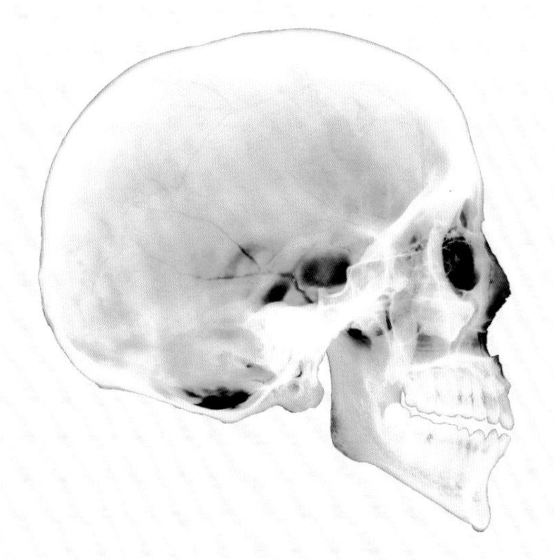

GORILLA
Average volume (males) 500–540ml
Average volume (females) 420–460ml
(from 240 ml in some very small females to 700 ml-plus in some largest males)

CHIMPANZEE
Average volume (males) 320–370ml
Average volume (females) 320–370ml
(260ml in some female bonobos to 420-ml plus in some large male common chimpanzees)

HUMAN
Average volume (males) 1,400–1,470ml
Average volume (females) 1,290–1,350ml
(from 1,050ml in some small females to 1,850ml in some large males)

Imaging) allow us to look, 'live' and completely without harm, at the regions of the brain which are most active when doing certain tasks such as speaking, listening, puzzle-solving, and manipulating objects. In one recent study humans and chimpanzees were brain-scanned while they were at rest, with no particular task to achieve. In this state, humans often think in an abstract manner about objects and situations elsewhere, perhaps planning ahead or assessing what has already passed. These scans showed many similarities between the humans and the chimpanzees in the study. In both species at rest there was plenty of activity in the parts called the prefrontal lobes, which lie at the front of the cerebrum. This suggests that chimpanzees, like us, have abstract thoughts when there is not much else to do.

During these scans the areas or centres of the human brain that deal with language – which are mainly located on the left side of the left hemisphere and include Broca's area, which deals with producing speech, and Wernicke's area, where spoken words are analyzed and understood – 'lit up' well. The chimpanzee has equivalent parts of the brain where calls may originate and be understood: these areas were not especially active during the study. However, another scanning session found that among chimpanzees, bonobos, and gorillas, there are other dominant areas in the left hemisphere that light up when the body is at rest. These are not the dominant left areas in humans that control language and speech. These findings feed a conundrum that scientists have long pondered. Many of our human thoughts are expressed mentally in words. Our language centres 'light up' as we think. However, chimpanzees and other apes do not have words, so how do they think? It is possible that their thoughts are composed of images. If not, then they think in some way we have yet to discover.

Multisensory
Sight for the target item, hearing for possible twig snaps, touch to secure grip, and balance to detect sway if the support gives way, mean orangutan 'fishing' is a complex task needing immense sensory and motor-skill synchronization.

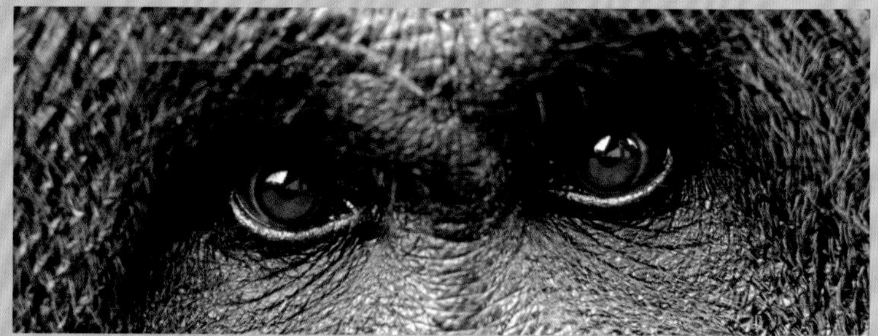

Senses

Looking into the eyes of an ape who is looking back can be an unsettling experience. As a family, great apes are predominantly visual creatures. We all conduct our lives mainly through sight. It is estimated that two-thirds of the knowledge and memories in the human brain came in through the eyes, in the form of scenes, images, written words, and diagrams.

Why do all the apes rely on sight so much? Part of the answer is an arboreal lifestyle. Primates in general are highly adapted to tree dwelling. When moving among branches, vision is tremendously important compared to the other senses – it is difficult to hear or smell your way through the tangled canopy. The importance of vision explains why most monkeys and all apes are diurnal (active by day). The only exception to this general rule are the douroucoulis or night-monkeys (also known as owl monkeys) of South America.

Two eyes on the front of the head, both looking forwards, results in a central overlap in their field of vision. This overlap allows the brain to compare the slightly differing views seen by the two eyes, a feature known as stereoscopic or binocular vision. Stereoscopic vision enables animals to perceive depth and judge distance accurately. Carnivorous mammals such as cats have it, to assess distances as they stalk and spring on their quarry. Primates also have stereoscopic vision, in their case to gauge accurately the distance of boughs and branches as they leap and swing through the trees. In many other mammals the eyes are positioned more towards the sides of the head. This is fine for providing an all-round view to detect predators and other dangers, but poor for distance perception (see page 126).

Windows on the world

The primate eye is well developed and relatively large compared with the skull and indeed the body size of its owner. Well protected in a bony bowl or orbit (socket) set within the skull, the eye has a domed clear 'window', the cornea, at its front. Light passes through this, then though the dark-looking hole just behind it, the pupil, which is set into a coloured ring of muscle, the iris. Behind the pupil is the lens, which changes shape to focus the light rays onto the inside lining of the rear of the eyeball, the retina. In the retina are many millions of microscopic cells, named rods and cones after their shapes. These cells respond to light shining on them by generating nerve signals that pass along the optic nerve to the brain.

It is difficult for us to imagine a world without colour. Primates have better colour vision than most other animals, and this is due to the large numbers of cone cells in their retinas. (Cones pick up colour, whereas rods are better at detecting light and process images in black and white. The world at night thus often appears to us only in shades of grey.) There are different kinds of cone cells with different light-responsive visual pigments for differing wavelengths – that is, colours (different colours result from different wavelengths of light). Some primates have two types of cone cells, but Old World monkeys and apes have three, sensitive to red, green, and blue light. Eyes with these three types of cone cells are known as trichromatic eyes. The two-plus million shades of colour we can distinguish result from different levels

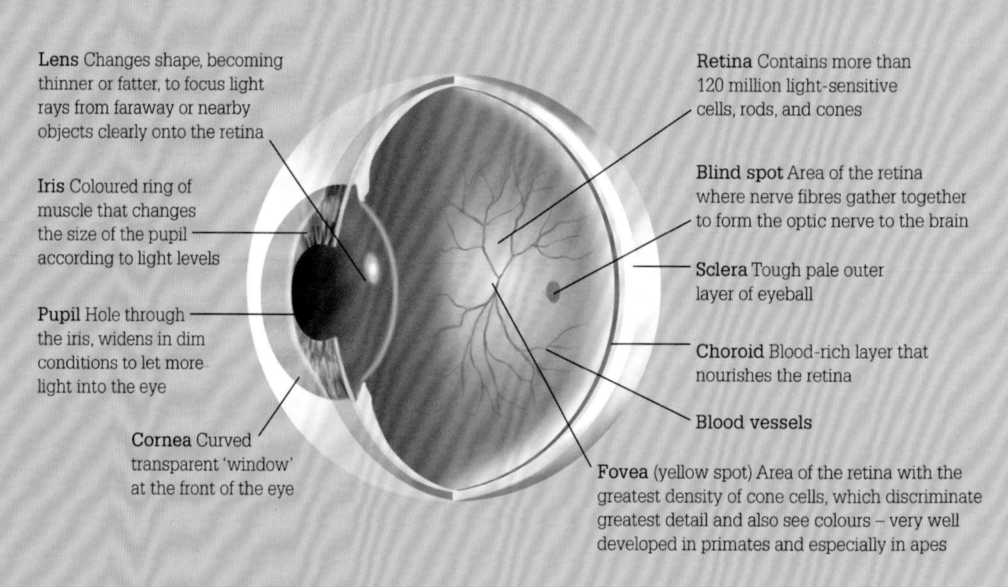

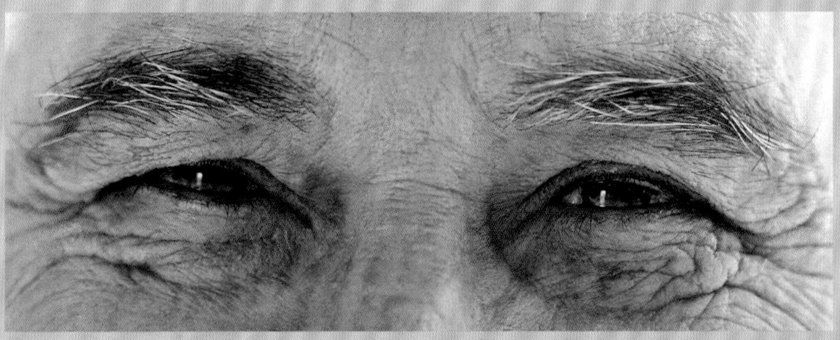

of stimulation of each of these cone cells. Animals with dichromatic eyes (containing just two types of cone cells) have poor colour discrimination, especially in the red-to-green region of the spectrum.

Eye to eye
All of the great apes have forward-facing eyes for detailed stereoscopic vision, but only human eyes show a range of iris colours. Above and opposite, left to right: Mountain gorilla, orangutan, chimpanzee, human.

The importance of colour vision

Ape eyes may not have the night vision capabilities of nocturnal creatures such as cats and owls, or the amazing telescope-like vision for faraway objects that birds of prey such as eagles have, but we have good visual acuity and advanced colour discrimination. Why is this the case? The traditional reason given has been that we need colour vision to choose the best foods, especially fruits. We can pick out precise shades and hues that tell us if a particular fruit is ripe and ready to eat, rather than unripe or over-ripe and so not worth bothering with. However recent research in Kibale National Park, Uganda has shown that such fruits can be equally well selected with dichromatic eyes. Leafy food, on the other hand, cannot. Trichromatic primates with red-green differentiation excel in spotting suitable bunches of juicy young leaves against the forest's complex background patterns of greens and browns. The younger leaves are often more slightly tinged with red than the tougher, more bitter older leaves. The ability to see this difference allows leaf-eating primates to select the most easily digestible leaves. Another factor that has been thrown into the colour vision debate in the past few years is sex. When humans blush or their skin flushes with extra blood flowing through the vessels just under the skin, we notice the reddening visually with our trichromatic eyes. The same happens when chimpanzees' rumps redden. Other apes have different sexual swellings to indicate arousal. So distinguishing delicate hues of red may help an ape to assess a partner's readiness to mate. The colour vision of more than 30 primates has been analyzed, and it has been found that those with most visual sensitivity to differing shades of red tend to be those that signal their state of sexual arousal via reddening of their bare rumps or faces.

Sacrificing smell

The possible link between keen colour perception and judging possible sexual partners has led to a further suggestion: that apes have sacrificed some of their sense of smell in order to improve their colour vision. (Most other mammals have poorer colour vision but a keen sense of smell, which they use to assess potential partners' readiness to mate.) Smell works by detecting airborne particles breathed into the nasal cavity, where two patches in its roof, the olfactory epithelia, bear millions of microscopic olfactory cells. Each of these cells has a tuft of hairs known as cilia. It is thought that different odour particles slot into different receptor sites on the cilia, like different keys into different locks, and thereby stimulate the cells to send nerve signals to the brain. All great apes routinely use smell to assess food for ripeness and palatability – orangutans, in particular, often hold and sniff items as they visually inspect them, before then testing their softness and flavour with the lips and tongue. The African apes, being more social, are still able to identify other group members by the odours of their urine, faeces, or glandular secretions, and can also determine sexual arousal, dominance, or threat using smell, at least to a limited degree.

However, primates in general and apes in particular do not excel at smell, especially compared to most other mammals. We have flattened snouts with smaller nasal cavities and fewer olfactory cells – an estimated 10–20 million in a human compared to more than 200 million in a bloodhound. The reason for our faded sense of smell may well be our improved eyesight. Apes rely less than other mammals on the sexually arousing scents known as pheromones, produced when individuals are ready to mate, because we have enhanced colour vision and can gauge the situation visually. Vision has the added advantage over smell of being multidirectional and able to operate over considerable distances. Detecting pheromone scents usually relies on close proximity or favourable wind direction. Support for this idea of smell sacrificed for vision comes from genetics. Analyzing the array of genes that carry instructions for olfactory receptor sites, researchers have found that in humans up to 60 per cent of the genes are pseudogenes – genes that are inactive, with their information not used for building receptors. This proportion is lower in the other great apes, around 30–40 per cent, and much lower again in other mammals, just 20 per cent in dogs. Superior sight may have decreased the need for an acute sense of smell, enabling apes to effectively switch unnecessary genes off and save the energy that would have been used in making superfluous olfactory receptor cells.

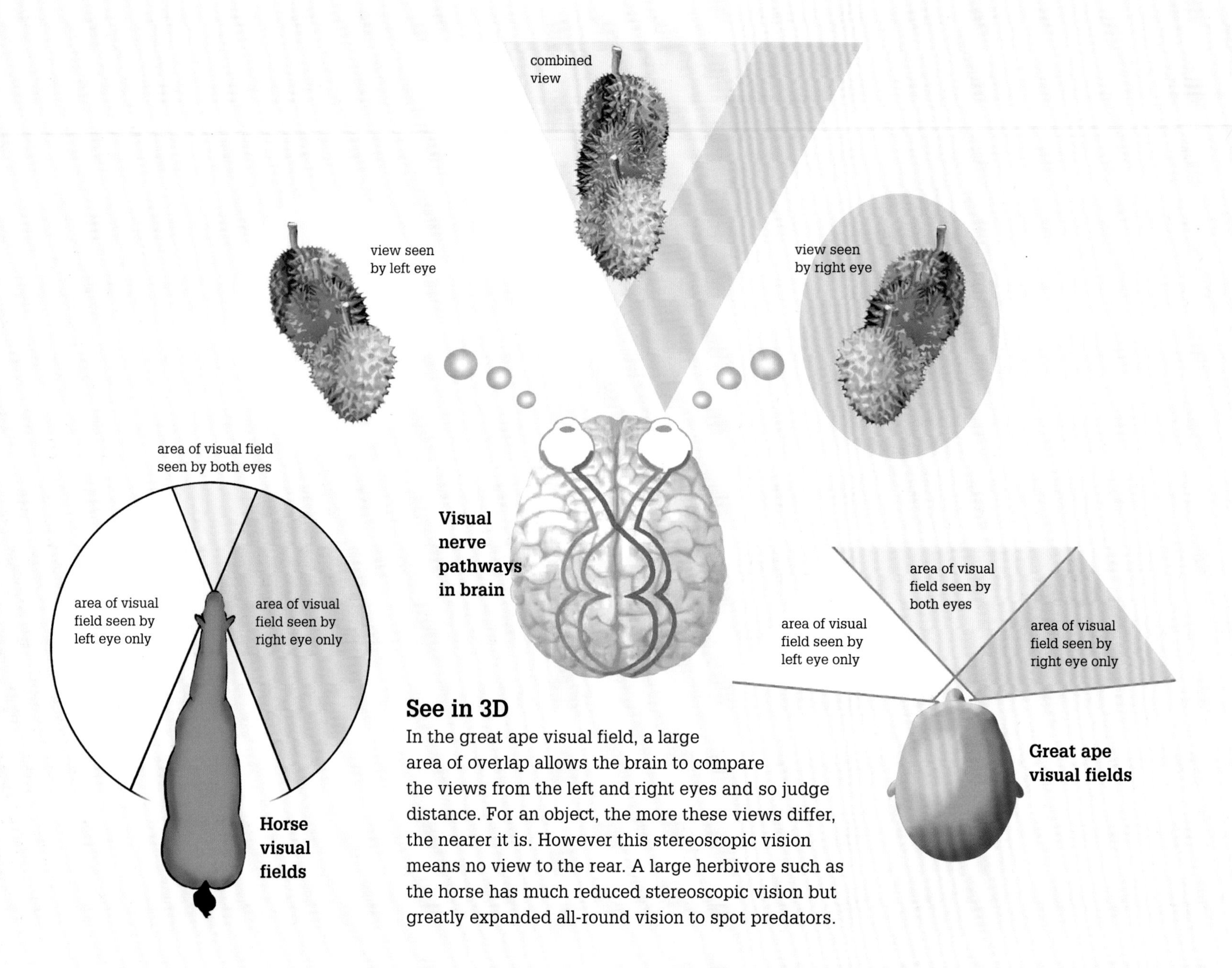

See in 3D
In the great ape visual field, a large area of overlap allows the brain to compare the views from the left and right eyes and so judge distance. For an object, the more these views differ, the nearer it is. However this stereoscopic vision means no view to the rear. A large herbivore such as the horse has much reduced stereoscopic vision but greatly expanded all-round vision to spot predators.

Hearing

The ears of all apes work in the same way. Sound waves are gathered by the outer ear flap, or pinna, and enter the tube-like ear canal. At the end of the canal they bounce off the eardrum, a thin disc of skin, causing it to vibrate. The vibrations are carried onwards into the head by a chain of three tiny bones, the ear ossicles. These are known individually from their shapes as the hammer, anvil, and stirrup. From the stirrup the vibrations pass into a fluid-filled organ shaped like a small snail shell, the cochlea. Here the vibrations shake the microscopic hairs (cilia) on rows of auditory hair cells, stimulating them to send nerve messages to the brain.

Pinna (ear flap) size varies among the apes from large and flaring in chimpanzees, to moderate in most humans, smaller in the gorillas, and even more reduced in the orangutans. In chimpanzees and humans especially there is great individual variation in pinna shape and pattern. The non-human apes live in forests and their visibility is often restricted by vegetation. They use their hearing to locate group members and other animals, and of course to detect danger. There are also differences within the ear between humans and the other great ape species. The shapes and leverage actions of our ossicle bones transmit vibrations more faithfully from eardrum to cochlea, preserving the delicate balance of frequency patterns, than those of other apes. However, in order to achieve this, the human design means that some energy is lost in transmission. This suggests that while we can hear in more detail, we cannot hear sounds at the lowest volume. Whispers or tiny rustles that might go undetected by us are audible to other apes. This is another trade-off. By sacrificing the ability to hear very quiet noises, we have gained the capacity to detect subtle differences within sounds, such as modulated speech. Another ability humans and most

Watchful eyes
Apart from a chimpanzee's own cautious sniffs and trial licks to check out unfamiliar foods, it also watches how others react, to see if they reject with aversion or tuck in with satisfaction.

other great apes have lost is the ability to move the pinnae, or outer ears. The muscles to do this, however, remain intact, evolutionary left-overs or vestiges with no biological function.

The mystery of taste

Perhaps the most pleasurable, but also the most mysterious, of the main senses is taste. Apes use it to assess foods for ripeness, bitterness, and possible contamination such as fungal growths. Taste is based in up to 10,000 small groups of cells known as taste buds, which are scattered mainly along the front, sides, and rear of the tongue. Each taste bud is a cluster of microscopic cells arranged like the segments of an orange, bearing cilia at their tips. As with smell, it is thought that flavour substances in foods fit into the receptor sites on the cilia, like keys into locks, thereby triggering the cells to fire off nerve messages to the brain's gustatory (taste) centres.

Familiar to many of us from school textbooks is the 'tongue map' illustration that shows which parts of the tongue detect the four basic flavours – sweet at the tip, salty and sour along the sides, and bitter at the rear. However, this map has suffered from uncritical perpetuation through the years and is now regarded as highly misleading. A fifth basic flavour has also been added to the four. That flavour is known as umami and perceived by us as a type of savouriness. Umami sensors are triggered by such foods as soy or miso enhanced with monosodium glutamate (MSG). Experiments with chimpanzees show that they can detect MSG in much the same way as humans. And it seems that while different gustatory cells are specialized to pick up only one or two flavours, these cells are widely distributed on the tongue. This means that most areas of the tongue can detect most of the basic flavours, rendering the old tongue map obsolete.

Internal organs

The other great apes have bellies very much like ours. The ape torso or trunk is a two-part repository for the soft internal organs, or viscera. The top compartment, known as the thorax or chest, contains the lungs, heart, and major blood vessels. The lower compartment, or abdomen, houses the digestive organs – stomach, ileum (small intestine), colon, and rectum (large intestine) – along with the excretory organs, being the paired kidneys and the bladder, and in the female, the reproductive parts such as ovaries and uterus (womb). The thorax and abdomen are separated by the diaphragm, an upward-curving sheet of muscle used for inflating and deflating the lungs.

The ape heart

Inside the chest, the ape heart beats steadily, pumping blood around the network of blood vessels. The muscular-walled heart is in fact two pumps working side by side. The left pump sends blood all around the body, from head to toes, along vessels known as arteries. The arteries divide repeatedly into tiny tubes thinner than hairs, called capillaries. Here exchange takes place, with oxygen and nutrients passing from the blood into the surrounding tissues, and waste products including carbon dioxide (CO_2) moving the other way. The dark-coloured, low-oxygen, high-CO_2 blood returns to the heart's right side, from where it is pumped on a much shorter journey to the lungs. Here it takes on board fresh oxygen supplies from the air through the lungs and unloads its CO_2. Now bright red, the blood then returns to the left side of the heart to continue its endless journey.

There is little difference in average heartbeat rates between the great ape species – in fact there is considerably more variation in the rates between individuals within species. In humans, most hearts pump between 60 and 80 times per minute, with averages for men of 70–74 and women of 72–76. Gorillas are much the same, with one study reporting a mean heart rate for male adults of 74 beats per minute. A wide-ranging survey of chimpanzee heart rates produced a healthy range of between 68 and 90 beats per minute. For all apes, the rate is faster in babies and infants. One newborn orangutan had a heart rate of 148 beats per minute: the human range at birth is 130 to 150.

Heartbeat rate increases during physical activity and also during times of stress, worry, or arousal – dominance disputes or sexual contact, for example. One of the most-studied chimpanzees ever was Enos, a US 'space-chimp' who in November 1961 became the second ape to go into orbit, soon after the more famous Ham. Enos's heart rate as he settled down in his capsule for blast-off was 94, with a breathing rate of 14 breaths per minute and a body temperature of 36.5°C (98°F). He was far more relaxed, normal, and 'cool' than most human astronauts.

Vege-primate
The gorilla's diet includes plenty of low-quality plant matter such as bark (as here) and stringy stems. A 'large' large intestine is needed to ferment this tough material and extract nourishment.

Big belly
A gorilla's full stomach and intestines can give a 'pot-bellied' appearance.

Heart rate is a useful parameter to measure when studying captive apes and their reactions to various situations and stimuli. It is non-invasive, and with careful training the subjects soon become used to the equipment. Chimpanzees shown photographs of other chimpanzees have been carefully monitored and their heart rate taken, for example, to gauge their level of arousal. Photographs of a familiar chimpanzee, a strange one, and an aggressive individual have been shown to produce different heartbeat rates, suggesting the chimpanzees recognize some of the differences in the images.

The blood group system known as ABO is present in all the great apes, including humans. Most humans are either A, B, AB or O, but there are other systems, such as the Rhesus (Rh) system present in our species. Other great ape species have their own rare blood systems, which occur alongside the more common ABO.

Breath of life

While the heart and blood systems are extremely similar in all great apes, including our own species, the lungs and respiratory systems show several minor differences. Human lungs are subdivided into parts called lobes, two in the left lung and three in the right. Chimpanzee lungs have a similar layout, but there is a difference at the next level of airway division. Humans have 11 main branches in the right lung and 10 in the left, whereas chimpanzees have 12 and 13 respectively. Gorillas have a two-lobed left lung but four lobes on the right. Each orangutan lung, left and right, appears externally to be composed of just one lobe. Despite their structural differences, overall, the great apes have similar lung capacities relative to body size.

One notable difference between the airway systems of humans and non-human great apes is that they possess (and we lack) large laryngeal air sacs. These air-filled chambers extend under the skin and muscle wall along the front of the neck, towards the collarbone area and down towards the armpits. In orangutans and male gorillas they reach all the way down into the chest wall. The laryngeal air sacs are connected to the windpipe by paired channel-like openings that occur just above the vocal cords in the larynx (voice box).

Laryngeal air sacs can be inflated or deflated by muscle action, and they have a range of uses, especially in vocal communication. Adult male orangutans have the largest air sacs of all, with a volume of up to six litres – the same as adult human lungs. They use their air sacs as resonators, to amplify the volume of the 'long calls' that advertise their presence to potential rivals. Bonobos use their air sacs in a similar way, to increase the volume of their low-frequency hoots. The hollow, echoing, 'popping' sound of a male gorilla chest-beat is also due to the sonic properties of air sacs. There is even a medical condition that affects laryngeal air sacs, known, perhaps predictably, as airsacculitis. Its symptoms are wheezy breathing, coughing, and a runny nose.

Air sacs are not confined to the great apes but occur in many other primates too. In recent decades there has been a great deal of debate over whether all the other great apes possessed them because they were present in our now long-extinct common ancestor. If they were then humans gradually lost their air sacs in recent evolution – perhaps because of our developing ability for speech. Recent research indicates that there are actually vestiges of air sacs in the human larynx. These vestiges have been given the name laryngeal ventricles. The lack of air sacs makes our voices much quieter but more amenable to fine manipulation for detailed sounds. Air sacs remained large in the other apes because they continued to confer advantages. As well as making vocalizations louder they can also be inflated to make their owners look larger and more intimidating.

Intestines and digestion

The ape digestive tract has the same basic organization in all species. Food is swallowed down the oesophagus, or gullet, and then enters the stomach. Here it is treated to a corrosive bath of hydrochloric acid and enzymes – chemicals that hasten its digestion, the process by which it is broken down to release its nutrients.

After a few hours festering and fermenting in the stomach, the partly digested food trickles into the long coils of the ileum, or small intestine. Here more enzymes are added to continue its breakdown, and the resulting nutrients are absorbed into the blood flowing through the intestinal wall. At the end of the small intestine is a chamber, the caecum, which has a finger-like projection known as the appendix. All great apes have an appendix, but this organ is less developed or completely absent in all other primates.

From the caecum, the remaining matter enters the colon – the first part of the large intestine. The large intestine is so named because it has a greater diameter than the small intestine, however it is shorter in length. In the colon water, useful salts, and minerals are absorbed. The leftovers are squeezed and packaged as a brown mass, the faeces, ready for elimination through the final parts of the tract, the rectum and anus.

Gut reaction

The similarity between great ape digestive systems was nicely illustrated in a recent news story. Visitors to a medical conference were asked to view two images of intestines. The doctors, surgeons, and other onlookers all pointed out various interesting details. But none of them noticed that one set of intestines came from a human and the other from a gorilla.

Despite these so-called experts' inability to tell the two apart, there are some differences in the relative proportions of human and other great ape digestive systems. Compared to our own digestive system, the other apes have a relatively larger large intestine and smaller small intestine. In humans, the small intestine contributes some 60 per cent of the total gut volume, while in chimpanzees and orangutans, for example, it makes up only 20 per cent. The human large intestine, on the other hand, makes up just 20 per cent of the total gut volume, while in the other great apes this figure is more than 50 per cent. The one area where humans and the other great apes coincide is the stomach. In all great ape species, our own included, the stomach forms 15–20 per cent of the total volume of the digestive tract.

The reason for the difference in the relative sizes of our large and small intestines compared with those of the other great apes is diet. The other great apes eat far more plant matter than most human beings do. Plants are much more difficult to digest than animal matter. As a consequence you need to eat more, and give it more time to break down and release its nutrients. This is where a bigger large intestine comes into its own – especially one teeming with billions of 'friendly' microbes. These microbes – mainly bacteria – make up what is known as the gut flora. In recent years many of us have become aware of 'probiotic' drinks, yogurts, and similar products that promise to keep our gut flora healthy and our intestines working effectively. For the other apes the gut flora plays a much more important role than it does in our own species. Their gut flora are more abundant and more effective at breaking down tough plant material in the colon to extract maximum goodness. Gorillas in particular rely heavily on these microbes to survive.

Internal organs

The large intestine or colon is relatively bigger in the other great apes than in humans, while the small intestine or ileum is more reduced. The more herbivorous foods of the hairy great apes need longer to digest and ferment, enabling the breakdown and extraction of valuable nutrients from fibrous plant matter.

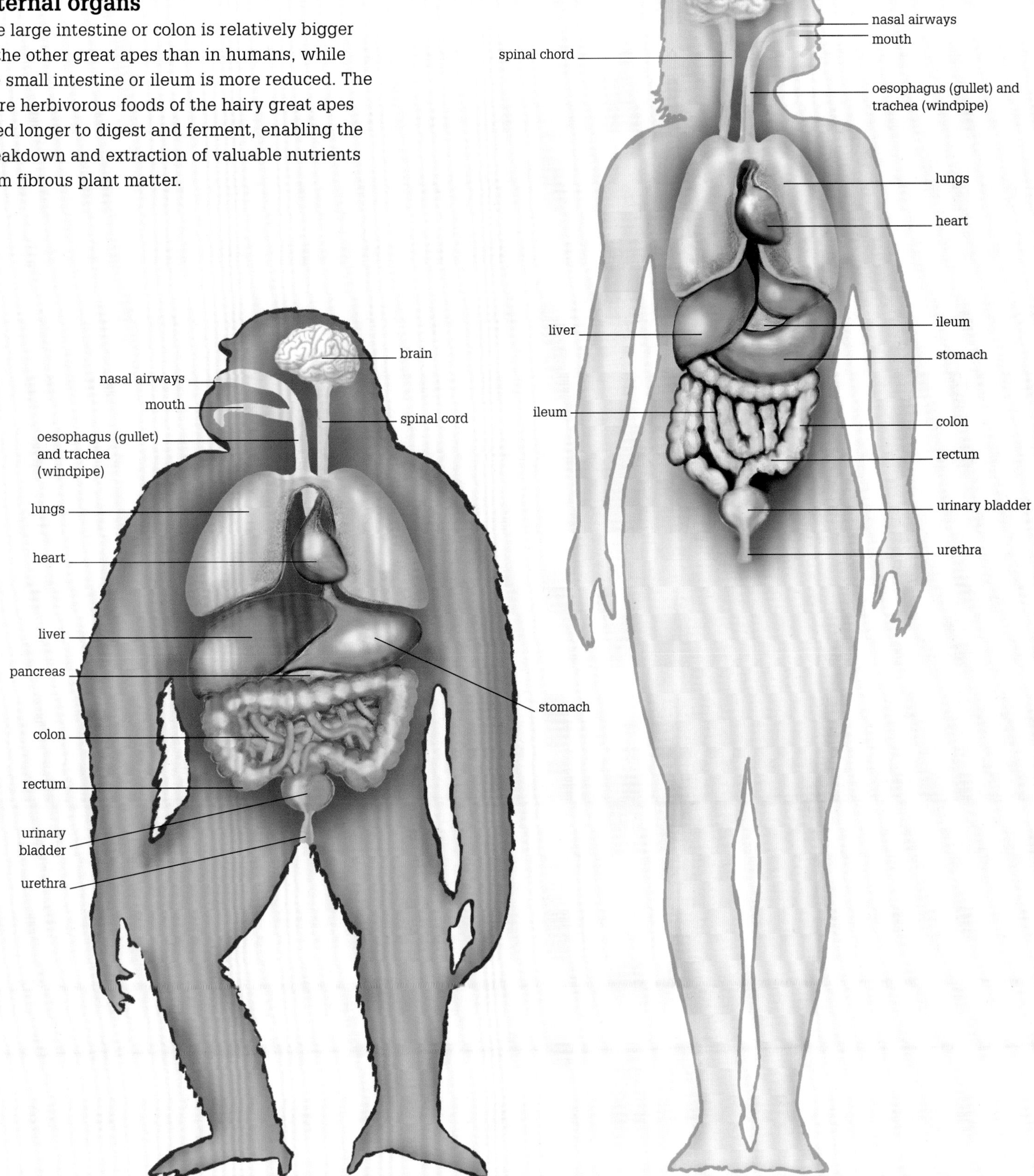

The Daily Meal

For affluent humans in today's convenience world, food experiences vary from grabbing a quick snack on the go to a supermarket dash for microwave-ready meals or savouring a lush feast at a restaurant. For our close cousins in the wild, food – in particular, finding it – dominates life in ways we cannot imagine. Finding enough to eat is the main task around which each day revolves. But it's not always a daily grind. In times of plenty the great apes, like ourselves, take time to relish tastes and flavours. And for chimpanzees and bonobos, especially, food and its associated desires are often bound up with sexual favours and power within the troop.

Feeding and tool-use

People once believed tool use was an exclusively human trait and no animal could be clever enough to use objects like we do. But as we learned more about the animal world, the list of creatures that use tools in the wild grew, from Egyptian vultures and woodpecker finches to various monkeys and apes. Once these tool-users were discovered we decided that fashioning tools before use was what set us apart, but we were wrong again. Chimpanzees are not only the most inventive, adept, and varied tool-users in the animal world, they also make tools by modifying natural objects. The way chimps and other great apes make and use tools shows them to be highly intelligent: able to reason and plan in order to solve problems.

What are tools?

Tools are objects, either unaltered or modified, that are used as intermediaries in a task. Humans use tools for a great variety of purposes, and the tools themselves are extremely varied as a consequence. Animals use tools primarily to aid survival, and the majority are employed to help obtain food. That said, among the great apes inanimate objects can be used for display – broken off branches may be waved threateningly, for example. They can also be used for hygiene: some apes use leaves to wipe mucus, blood, sap, excrement and other unwanted substances from the body.

The chimp toolkit

The longest list of tools made and used by any of the non-human great apes is that of the chimpanzee. Among the best-known chimp tools are the various forms of rods or wands used for 'fishing' for termites, ants, and other colonial insects. The rod may be a ready-made twig, stem, or grass blade, used as it comes from nature. On the other hand, it might be modified, for example, by stripping leaves from a central stem. A termite 'fishing rod' is probed and poked into a nest opening, causing the termites to attack it. Once covered with the insects the rod is withdrawn so the chimpanzee can lick them off. Before use, a fishing rod is carefully selected for length, width and flexibility. The chimp may also modify the rod's business end by chewing it or pulling it between the teeth, to fray the edges into a type of brush, which is better at gathering termites or ants.

A simpler, more physical technique for obtaining ants or termites as food is to break open the nest with a big puncturing stick. This process is far from delicate, and it can only be used successfully when the nest walls are thin enough to be broken. Another technique, intermediate between puncturing and fishing, is to use a stiff but narrow perforating stick, which is poked or drilled into the nest. As the termites or ants swarm onto the stick, which is held in one hand, the chimpanzee swipes them off with its other hand and stuffs them into its mouth.

Recent research on chimpanzees in the Nouabalé-Ndoki National Park, in the Democratic Republic of Congo, has shown how sophisticated tool-use can be. The chimps here use a particular type of stick, usually from a small tree known as *Thomandersia hensii*, to break open termite nests. To probe deeper and fish out the termites they employ stems from a different plant, the herb *Sarcophrynium*. Youngsters watch their mothers and other adults carefully, noting how, when feeding, they avoid the large soldier termites, which have powerful jaws and painful bites. These chimpanzees not only make tools on site, but may arrive at termite mounds with ready-made sticks and probes, prepared for the task ahead. This behaviour, which has also been observed in Sumatran orangutans, demonstrates conscious planning for future events – another trait that we humans once fondly imagined was exclusive to ourselves.

Hammers and spears

Chimpanzees get into palm oil nuts, baobab nuts and similar hard-cased foods by using a makeshift hammer and anvil. The anvil takes the shape of a flat stone or exposed tree root. For the hammer, the chimpanzee selects a pebble or stone of suitable shape and weight, feeling it before use to check for grip and waving it to gauge its relative force and momentum. It then makes several test blows, gradually reducing their force to save energy and avoid injury, while ensuring they remain powerful enough for the job. The best anvils close to fruiting tree are much sought after, and at times chimpanzees actually queue up to use them. Some chimpanzees can spend up to three hours a day waiting and then cracking open nuts. An alternative strategy is used for larger nuts, such as coconuts, which are smashed directly onto the anvils.

Sticks poked into tree hollows can reveal wild bee colonies, and then be used for honey dipping, or they can be used to locate ant or termite nests. They can also be used to disturb small sheltering

Tuber time
(Previous page) Thickened food-storing parts of plant stems and roots, known as tubers, provide this male mountain gorilla with tasty, carbohydrate-rich, energy-giving snacks.

Test tasting
A chimpanzee tastes a few small grubs from a nest, which have adhered to the end of its 'fishing stick'.

Careful positioning
Using a log as the base or 'anvil', a chimpanzee positions a nut, ready to crack it open with a 'hammer' pebble carefully selected for size, weight and grip-able shape.

Fishing for termites
A chimpanzee probes hopefully into a small opening through the dried-mud, rock-hard wall of a termite mound. This may provoke the inhabitants to 'attack' the intruding stick, and be withdrawn to their doom.

animals, which are then jabbed and poked out of their hiding places for consumption. This jabbing or spearing of small animals such as bushbabies in tree holes is a relatively recent discovery. The 'spear' may be an unmodified stick or broken-off branch, but some spears are adjusted by stripping the bark and even sharpening the end with the teeth. The pointed end is examined and sniffed after each series of prods to detect telltale signs such as hair, skin or blood that would give away an occupant. The intention is to jab forcefully enough to injure the small animal, which then tries to escape and is captured and eaten.

Some chimpanzee habitats include regions of dry bush, where water is scarce at some times of year. Chimpanzees in these places use various types of leaves as sponges for gathering drinking water from tree hollows and other relatively inaccessible spots. A leaf may be modified to make it more absorbent and porous, by folding and then chewing it or rubbing it on a stone or branch. Leaves of *Hybophrynium braunium*, from the arrowroot family, and those of the andiroba tree are among the most commonly used.

Learning by example

Watching a parent or other adult use tools is an important part of learning for young chimpanzees. Youngsters study how adults fish for termites, use leaf sponges, or hammer nuts, and gradually try to mimic the behaviour. In some instances, mother chimpanzees have been observed actively teaching or coaching their offspring. Again, this is a trait that was once believed to be solely human. In one case, as a young chimpanzee struggled to open a panda nut, the mother showed how to reposition the nut on the anvil so that it cracked more easily. Another mother was observed showing her infant how to hold the

hammer for best effect. It is also known that different groups of chimpanzees use their tools in different ways.The method of using a particular tool, learnt by imitation, is passed from one generation to the next, and is exclusive to that group of chimpanzees. As a result, making and using tools has been cited as evidence for material culture among these apes.

Quenching thirst
Rainwater collected in a hollow log or stump may be out of mouth's reach.A squashed and crumpled leaf works well as a sponge to soak it up and squeeze out a refreshing drink.

My favourite stone
Côte d'Ivoire's Taï National Park has relatively few stones. So chimpanzees try to remember where they last used them for nut-cracking – much as a squirrel recalls where it buried nuts. Here a young chimpanzee sleeps peacefully through the adult's skilled demonstration.

Copy-cat apes
In the wild, Sumatran orangutans seem more adept at tool use compared to the Bornean species. In captivity both species readily manipulate and utilize many objects for play and purpose – possibly after observing humans doing similar activities. Here a pointed stick is employed to dig into soft bark, maybe for juicy grubs.

Teaching time
Adult chimpanzees do not only allow youngsters, like this juvenile male, to observe tool use passively. The 'teachers' have been seen actively to demonstrate the correct technique and help their 'pupils'.

Tool-use among other apes

Perhaps surprisingly, given the great range of tools used by chimpanzees, bonobos show very little tool use in the wild. This is in marked contrast to bonobos in captivity, which readily employ fishing sticks, hammer stones, and other items when given the opportunity. The same dichotomy is seen among gorillas. In captivity they often use tools, but they have almost never been observed using them in nature. (The exception is the use of 'walking sticks' to test the depth of water in swamps.) In all probability, wild gorillas fail to use tools simply because they do not need them, living, as they do, surrounded by an abundance of food.

Like bonobos, Bornean orangutans quickly learn to use various tools in captivity, but evidence for tool-use in the wild is scarce. The only objects from the environment regularly used are large leaves, which are held over the head as sunshades or umbrellas. Leafy branches above tree nests are also sometimes woven together to create a roof. The situation is very different for Sumatran orangutans, which have been seen using sticks in the wild in several ways. Like chimpanzees, Sumatran orangutans fish termites and ants from their nests, by pushing sticks into holes in nest walls. They also use sticks to extract seeds from their protective pods or husks. Before use, the sticks are often modified by being broken or rubbed into shape, or having their leaves removed. This shows a level of intelligence that humans, 50 years ago, could never have imagined of their great ape cousins.

Nutritious nuts
Nuts and hard-cased seeds are packed with nourishment but are well protected. Activities such as nut-cracking must pay a high enough nutrient yield in terms of learning and practice, finding the tools and nuts, and actually cracking them open, to offset the time and effort spent in these processes.

Tasty termites
Soft-bodied, full of dietary goodness, and largely defenceless – because of the hard-walled nest around them – the worker termites fished out with a prepared twig make a rewarding meal. The soldier termites, with their powerful sharp mandibles, need more care in consumption.

Essentials of diet

Well worth the effort
For this male mountain gorilla, the laborious efforts of stripping and chewing bark are rewarded by colonic digestion of the cellulose and other bark contents – with the bonus of a quick lick of juicy, oozing sap.

The vast amount of information we have about the human diet and health is partly transferable to the apes. We know from centuries of studies on our own nutrient intake that lack of protein, vitamins, or minerals can bring on disorders. Scurvy, for example, is caused by a lack of vitamin C, while the pot-bellied signs of general malnutrition result from a shortage of protein. Increasingly, studies on captive apes have revealed how their bodies react to different kinds of diet, including the problems of excess when orangutans, gorillas, and chimpanzees are fed too much rich food and take too little exercise.

Carbohydrates

The chief sources of energy in most foods are carbohydrates, also known as sugars and starches. In our own diet they tend to come mainly from grains such as wheat, maize, and rice, and their products, as well as from vegetables such as potatoes. Among apes, the main sources of carbohydrates are high-starch seeds, nuts, and plant storage parts, such as tubers, pithy stems, and bulbs. Starchy foods are broken down by digestion into a mix of simpler sugars, the most important of which is glucose, which travels around in the blood and is the main energy source for most of the body's life processes. Sugars are also consumed directly in the shape of honey, and in the tissues of ripe fruits and the bodies of insects and other animal prey.

Proteins

Most of the body's structural components are built up from a vast array of proteins. Proteins are the building blocks of everything from bones

Typical dietary intakes

orangutan

50–70%
fruit (durian, fig, lychee, neesia, rambutan, kamala

20–30%
leaves, shoots, stems, seeds, buds, bark, and roots

2%
small animal items

western lowland gorillas

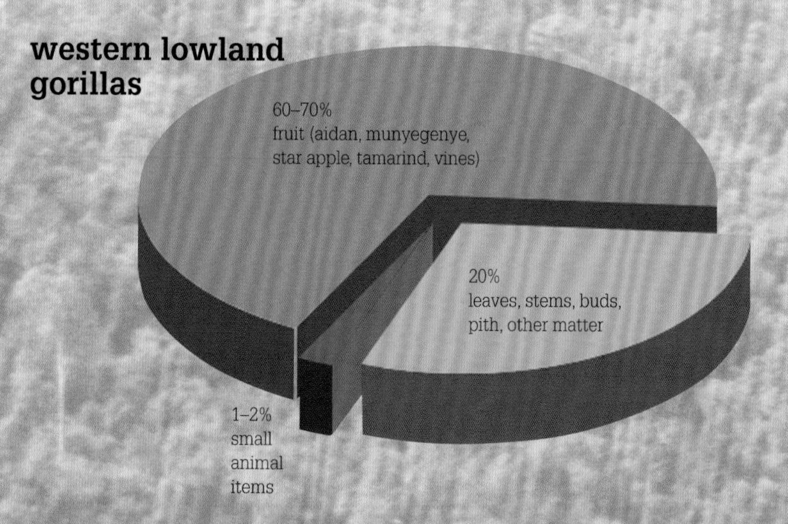

and teeth to cartilage (gristle), sinews, and muscle. To grow, maintain, and repair these body parts requires a certain amount of protein intake, both from plant and animal sources. Like carbohydrates, proteins can also be broken down into simpler sugars to provide for some of an ape's energy needs. Despite what is often thought, all of the great apes consume a certain amount of animal protein. Some do this intentionally, seeking out eggs, chicks, grubs, ants, and larger animal prey. Others do it inadvertently, swallowing insects and other tiny creatures as they munch through leaves, roots, stems, and fruit. The importance of protein in the ape diet cannot be underestimated. In one field study of western lowland gorillas in Gabon, protein deficiency was shown to lead to hair loss and discoloration, weight loss, digestive upsets, and even death. Gorillas that were offered protein supplements (intended for humans) rapidly returned to full health.

Fats

Fats and oils are essential in a mixed diet for general health. They are also important for the role they play in energy provision and for the formation of certain body parts, such as nerves. The overconsumption of animal fats can cause health problems, such as increased risk of heart disease, but this is not really an issue for great apes, at least in the wild. Most of the fats that they consume are the so-called polyunsaturates found in plant matter which promote health rather than have a negative impact on it. Their animal fat intake is minuscule compared to ours.

Vitamins and minerals

Wild apes rarely suffer from vitamin or mineral deficiencies. Their natural instincts encourage them to eat all manner of items that seem unpalatable to us, including soil, mineral-rich salt deposits, rotting wood, and even animal droppings in order to obtain a full and varied range of vitamins and minerals. However, chimpanzees and orangutans in captivity have been known to suffer from vitamin D deficiency when fed a limited diet and deprived of sunlight. Like humans in the same conditions, these apes have gone on to develop childhood rickets or adult osteomalacia.

A tough nut to crack?
For the young orangutan, youthful inquisitiveness (in addition to observing mother) is an essential part of discovering what is good to eat – nuts and seeds are rich in proteins and fats essential for growth.

Fibre (roughage)

The microscopic cell walls of plants are made of a material called cellulose, and this is the major source of dietary fibre. In humans it reduces appetite and eases the passage of food through the digestive tract. It also protects against certain diseases, such as intestinal cancers. Similar effects are seen in apes. However, ape guts treat fibre differently from our own. In gorillas, especially, the colon contains certain bacteria that ferment a proportion of fibre into fatty substances, which can then be used as a source of energy.

Due to their predominantly vegetarian habits, all great apes consume far more fibre than the average human. By dry weight a typical gorilla diet is 40–45 per cent fibre and a chimpanzee's is around 30–35 per cent. In a typical European human who is not that diet-conscious, it is just 5 per cent.

mountain gorilla

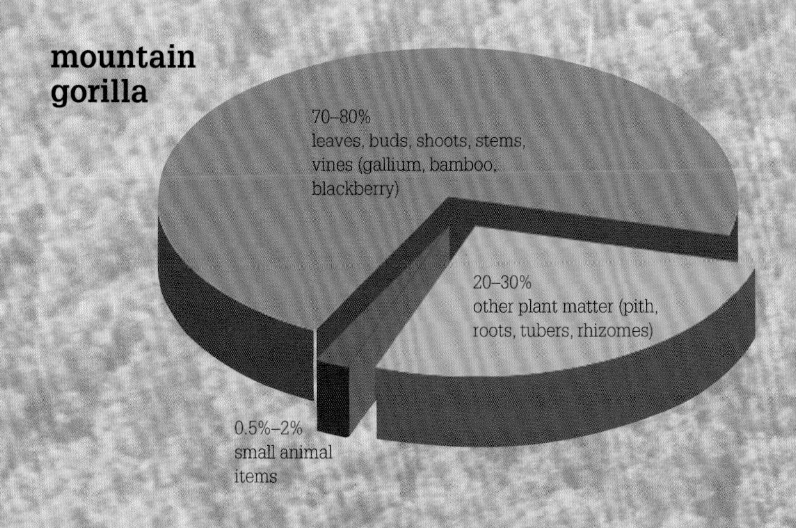

chimpanzee

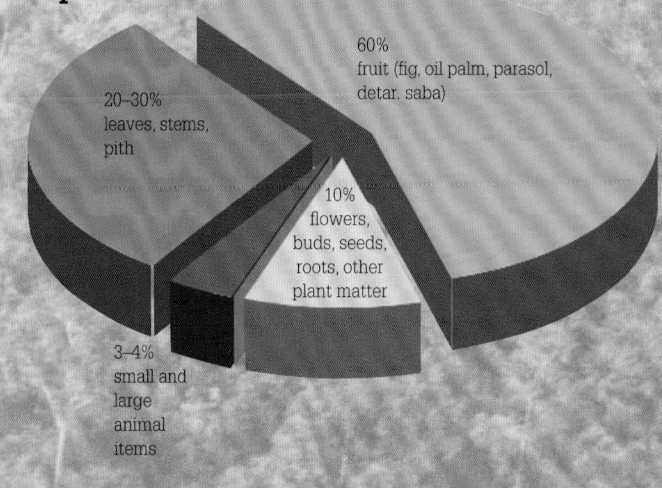

Daily needs

Quality vs quantity

In terms of volume, gorillas, not surprisingly, consume the most food, and with their active lifestyle and brain function, it is not surprising that the chimpanzee follows. The orangutan's leisurely lifestyle means they consume less in one sitting, though as with all apes, their day revolves around food.

ORANGUTANS
2 durian fruit, 4 figs, 4 lychees
miscellaneous mixed green leaves
miscellaneous shoots and flower buds
shreds of bark
2 eggs

GORILLAS
bucketful miscellaneous mixed green leaves
bunch long green bamboo stems
bunch wild celery
3 A5 slabs of bark
length of gallium vine including stem, leaves, flowers, and berries.

Quality and quantity
A gorilla eats roughly the equivalent of a bucketful at one sitting, and in one day will eat as much as 15 kilograms (33 lbs) a day. By comparison the human main meal (based on 3 meals a day) might be around a quarter of a bucketful (though as with all primates a human diet varies).

CHIMPANZEES
jamjar termites
fruit (e.g. durian fruit, figs)
½ bucketful miscellaneous mixed green leaves
½ bucketful flowers and fruits of the saba liana vine
lump wild bee honeycomb
½ jamjar mixed small insects
10 detarium nuts

HUMANS
bread
100g grilled chicken
50g vegetables
50g rice
fruit
milk, butter, and yoghurt

Diet and dentition

Display of weaponry
Apart from various eating functions, teeth have a role in visual display, physical intimidation and attack. A threat face reveals the male gorilla's formidable canines by pulling the top lip up and back.

Knives, forks, and spoons help humans to process foods before they even enter the mouth. Indeed, in certain company it is considered impolite to carry out any kind of overt biting, even from an apple. Of course, non-human apes lack cutlery (although they do use some tools, as explained previously), so their mouths, teeth, and lips must carry out many more tasks relating to food than our own. These include slicing and ripping through tough foods, such as fibrous vegetation or meat.

Teeth and jaws

Humans and non-human apes all possess the same dental formula – we have an identical number of teeth, and their types and positions in the jaws are the same. In shorthand this formula is written I 2/2 C 1/1 P 2/2 M 3/3. In layman's terms, this means there are two incisor teeth, one canine, two premolars and three molars in each side of each jaw, upper and lower.

The outer layer of a tooth is made of enamel, which is by far the hardest substance in the body. However, an ape's lifetime of tough chewing gradually wears the enamel away. By the time a chimpanzee turns 40 years old its teeth usually show severe wear, which can lead to abscesses, gum disease, and tooth loss. Among gorillas, the

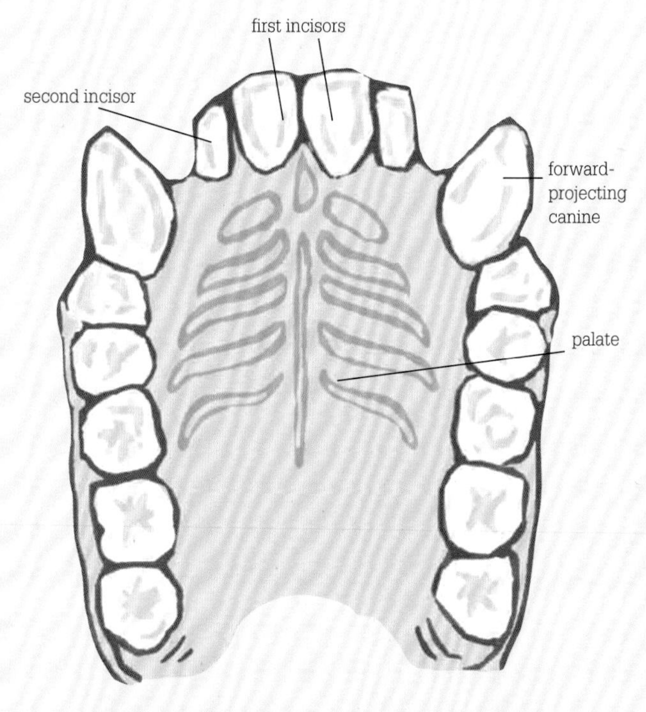

The orangutan's canines tend to project forwards more than in the other apes. Enamel thickness is intermediate between humans, with thick enamel, and the African great apes where it is thinner.

mountain gorilla has a diet lower in sugar than lowland gorillas, and studies show that it also has a lower level of tooth decay. Without a dentist, once teeth start to rot, infection can soon spread into the jawbone, sinuses and other nearby parts of the face and skull. Tooth infection and loss is a well-known cause of death in wild apes. Unable to chew food properly, they eventually starve to death.

Tasks for teeth

Each of the different types of tooth has a certain role when feeding. Incisors are straight-edged like small spades or chisels. They are ideal for nibbling and jobs such as removing outer fruit flesh from the hard seeds within. Canines are long, strong, pointed and slightly curved. Their major task is to rip into and tear up food.

Premolars and molars are broader and flatter, with a pattern of low mounds or cusps. They are the main chewing teeth, used for crushing and pulverizing. Humans tend to eat processed and cooked foods that are relatively soft and need little mastication. Great apes – particularly the mountain gorillas, whose diet is largely leaves – chew on average for more than 20 times longer each day than we do.

The role of canines

The human dental arch (the shape seen when the lower jaw is viewed from above, or the upper jaw is seen from below) is a rounded, smooth curve, tending towards an arc or 'U' shape. The dental arch of other apes is more like three sides of a rectangle, with a relatively sharp angle between the incisors at the front and the canines and chewing teeth along each side.

The other great apes have an appreciable gap, the diastema, between the second incisor and the canine, which is absent in humans. The diastema in the lower jaw accommodates the large upper canine when the mouth is closed, and vice versa for the upper jaw and lower canine. This is linked to the many roles of a great ape's canine teeth. In feeding, they are heavily used to strip away bark and reveal softer juicy sapwood, slice up stems to access pith and juices, rip into the softer interiors of hard fruits and shred leaves from twigs. In some individuals, deep grooves form on the upper incisors and notches on the lower canines as a result of stripping leaves from stems.

Away from feeding, the fearsome canines are used in threat displays against rivals, being exposed by pulling back the lips. They are also sometimes used as weapons, slashing and biting as the ape lunges at an adversary.

His and her teeth

In all non-human apes the male has larger canines, in proportion to the rest of the teeth and jaws, than the female does. This physical difference is linked to the importance of canines as weapons, as well as their role in display. Males with large canines are seen by females to be better able to defend themselves and their troop. As a result, they are more likely to be selected by females as mates. The male-female canine difference was evident in our own ancestors too, but it has disappeared over the last few million years of human evolution. Human canines are now not that much longer than our other teeth and there is little difference between the sexes. Exactly why the emphasis on canines became lost in humans is not clear.

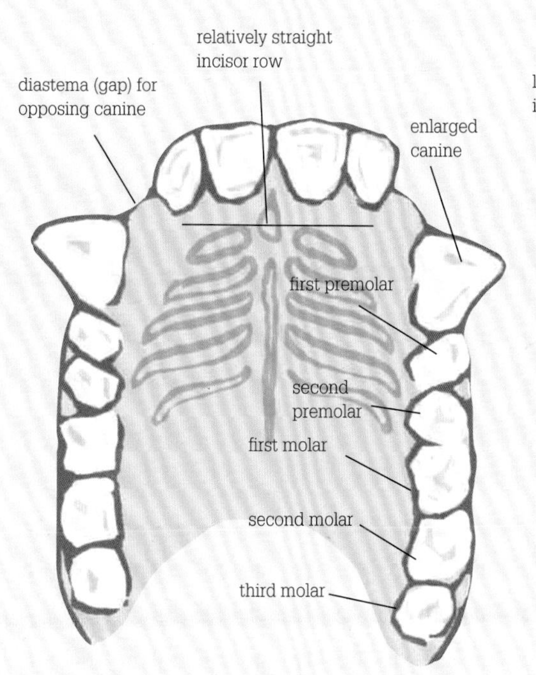

The four-incisor 'battery' at the front of the chimpanzee's mouth forms a relatively long and straight row, similar to the orangutan but not to the gorilla. The canines are partly angled outwards..

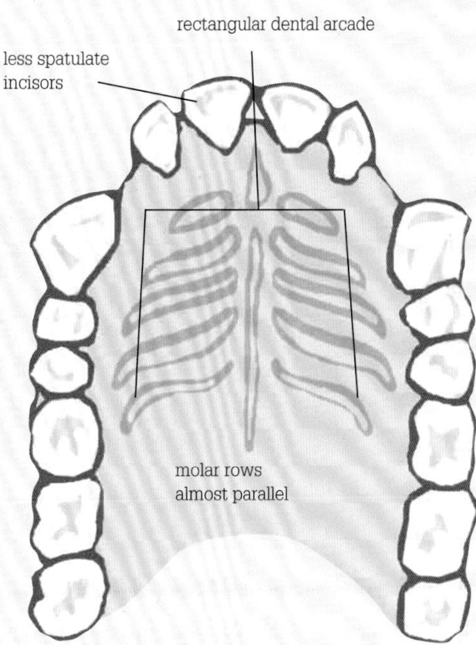

The gorilla's incisors are less spatulate, broader tipped and narrower based. This is probably linked to the reduced need for 'scooping' out the contents of fruits due to less overall fruit in the diet.

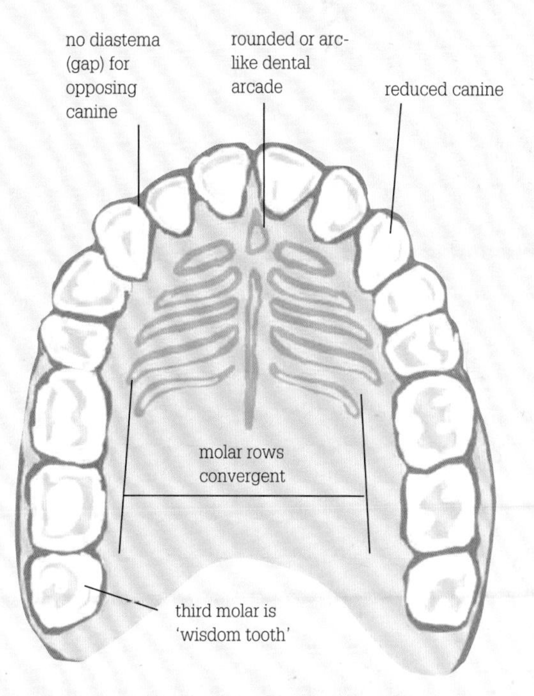

The canines of the human are much reduced and only slightly larger than the other teeth, and there is no diastema. The enamel layer is comparatively thicker, and the teeth form an arc-shaped dental arcade.

A life of plants

'Orangutans are frugivores, gorillas are folivores, chimpanzees are omnivores and humans tend to carnivory.' It's a handy saying but far from the whole picture. Gorillas do not restrict themselves to leaves, nor do orangutans exist solely on fruits. There is enormous variety in diet among the wild great apes, not only between the species, but between different populations of the same species. Diet varies through the seasons of the year, in bouts of extreme weather such as drought, between males and females, young and old, and dominant and submissive animals. Learned preferences and cultural differences also play their parts. What all great apes have in common when it comes to diet are their opportunism, adaptability, and intelligent use of food resources.

Basic preferences

Studies of orangutans identify fruits as the major single dietary component, comprising 50–70 per cent of their food intake. The remainder is a miscellany of leaves, shoots, stems, seeds, buds, bark, and roots, plus occasional small animal items such as termites, crickets, and other insects, eggs, small birds, and mammals.

Fruit makes up 60-70 per cent of the diet of western lowland gorillas. Leaves, stems, buds, and pith comprise most of the rest, with 1–2 per cent made up of small animal items. Mountain gorillas live largely on greenery, with leaves, buds, shoots, and stems making up 80 per cent of their diet. In their high-altitude habitat fruit is scarce.

Chimpanzees and bonobos have the most variable diets. On average roughly half of what they eat is fruit, 20–30 per cent is leaves and stems, 10 per cent is other plant parts such as flowers, buds, seeds and roots, and some 3–4 per cent is animal matter.

The orangutan diet

Orangutans have strong dexterous hands and powerful jaws that allow them to get into most of the fruits and other plant products that grow in the rainforest canopy. They have been recorded consuming fruits and other parts of more than 400 plant species. One of their favourite fruits is the durian, from the plant genus *Durio,* which has a spiny husk and soft flesh enclosing the seeds. The smell of durian would put most humans off eating it – its odour has been likened to that of excrement. Orangutans, however, are drawn towards it, sometimes gathering in numbers at fruiting durian trees, opening the husk with a combination of careful bites and manipulative fingers.

Another staple fruit of orangutans is the fig. As with durian, several species of fig are eaten. In some regions orangutans seem to avoid figs if other fruits are available, suggesting that figs are not necessarily a

favourite food. Lychees also feature in the orangutan diet, and their seed is often spat out with great ceremony. Other fruits eaten include jackfruit, breadfruit, and the fruits of many trees with only scientific names, among them *Nessia* (the seeds are extracted with tools), *Sarcotheca* (a cousin of wood sorrel), *Nephelium*, *Tetramerista*, *Mallotus*, *Gironniera*, *Lithocarpus*, *Antiaris*, *Tinomiscium* and *Eugenia*.

Flowers are sniffed cautiously and then the petals are plucked delicately with the lips. Those eaten include the blooms of *Xanthophyllum rufum*, known locally as minyak berok.

Young and juicy

Like other apes, orangutans prefer young, soft, juicy plant parts to older ones – especially leaves, since these develop tannins and other unpalatable chemicals as they grow, precisely for the purpose of

My favourite!
A young Bornean orangutan examines a durian fruit, looking for a weak spot among the thorns of the husk. Almost revered by orangutans, this fruit is known by people in the region both as 'king of fruits' and for its penetrating odour.

Eats, shoots, and leaves
The orangutan's 'four arms' mean there's always a way to reach and pluck a desirable leafy snack in almost any part of the forest.

Sight test
Good colour vision allows the orangutan to gauge initially whether leaves are young and soft rather than old and tough. This is followed up by feeling and sniffing to see if the handful is worth a chew.

repelling browsers. Leaves of various species in the breadfruit group, *Artocarpus*, are consumed with relish, as are those of trees of the genus *Baccaurea*, among others. Seeds feature here and there in the orangutan diet but are more commonly eaten by the Bornean than the Sumatran species. In Borneo many of the dominant rainforest trees belong to the so-called dipterocarp family. These trees fruit irregularly and produce extensive crops of two-winged, oil-rich seeds, which bring hungry Bornean orangutans swinging through the boughs.

The gorilla diet

Gorillas are the least frugivorous of the great apes. As has already been mentioned, mountain gorillas normally have little fruit available, so they focus on leaves, pithy stems, and other green plant matter for most of the year. Western lowland gorillas eat the most fruit of any gorillas, since their habitats provide a fairly plentiful supply almost year round. Eastern lowland gorillas eat less fruit than their western counterparts but more than mountain gorillas do.

Much of the fruit eaten by gorillas is collected on the ground. Fallen fruits favoured by western lowland gorillas include those of the aidan tree, or munyegenye, the black tamarind and the star apple tree, *Chrysophyllum cainito*. The fruits of various trailing vines are also eaten, as are numerous herbs. In the forest clearings, known as baïs western lowland gorillas often sit contentedly in soft mud, munching on aquatic plants such as sedges and marantochloa herbs. When fruits are unavailable or hard to come by these gorillas turn more to pithy stems and bark.

Eastern lowland gorillas have a more varied diet in general. High on the menu are various types of bamboo, wild celery (*Peucedanum linderi*), the bark of many trees, such as African redwood, and herbaceous plants such as members of the groundsel and ragwort groups.

Mountain gorillas have much less fruit available than their lowland counterparts, and their leaf-and-stem staples are also less varied. Even so, some troops studied have been found to exploit more than 40 plant species through the year. Near the top of the menu is the gallium vine, *Galium rewenzoriense*. This is consumed almost in its entirety, leaves, flowers, fruits, and stems, and is so popular that it has become known to researchers as 'gorilla fast food'. Other favourites are the mountain bamboo *Arundaria alpina*, particularly the soft young shoots. The stems and leaves of wild celery, bush morning glory (*Ipomoea leptophylla*) and Malabar spinach (*Basella alba*) are also popular, as are the young stems and roots of the stinging nettle *Urticaria massaica*. The few fruits that are eaten are mainly of various species related to blackberries and raspberries, such as the Ruwenzori blackberry (*Rubus ruwenssori*). The Bwindi population of mountain gorillas eats a higher

Chewing time
The gorilla's massive masticatory muscles work for many hours daily, pulping vegetation almost into a soup as part of physical digestion, so that it is easier for the stomach to work on by chemical digestion with acids and enzymes.

proportion of fruits than the Virunga population, but overall, fruits rarely constitute more than 20 per cent of the total diet.

The chimpanzee diet

Chimpanzees are omnivorous and exceptionally opportunistic. A full list of their dietary items would stretch the combined restaurants of a big city. Studies have shown that more than 320 species of plants are consumed (chimpanzees also eat a wide variety of animal matter), with an average of 50 different plants eaten every month. Sometimes, parts of as many as 20 different plants will be consumed in a single day. The variety of plants eaten by chimpanzees partly reflects the range of different habitats they occupy, from dry bush to swampy thick forest. Favourite plant foods for most chimpanzees include figs and the fruits of the saba liana vine (*Saba florida*), the oil palm (*Elaeis guineensis*), and the parasol tree (*Musanga cecropioides*). In some localities, the latter fruits almost all year round. Nuts of various trees are cracked open and eaten, including those of the sweet detar (*Detarium senegalense*), mubura (*Parinari excelsa*), and gere-gere (*Panda oleosa*) trees. An intriguing chimpanzee habit is that of 'wadging' certain foods, which might be compared to our own habit of extracting the juice from soft fruits. The food items, usually mixed leaves and fruits, are pressed and squashed in the mouth against the teeth and palate while the chimpanzee sucks hard to extract the oozing liquids. This process may continue for up to 10 minutes before the spent fibrous remnants are finally spat out. Wadging helps to avoid swallowing this tough, indigestible matter and also reduces the risk of biting into distasteful seeds that may be lurking in the mouthful.

The bonobo diet

Bonobos have a wide-ranging diet, although not quite as varied as that of chimpanzees. Around half of their food intake is made up of fruits, including figs and other items generally similar to those eaten by chimpanzees. The huge fruits of the tahu tree (*Anonidium mannii*) occasionally provide a windfall meal as they crash to the ground. Those of the African breadfruit (*Treculia africana*) are even larger, weighing as much as a bonobo itself, but they are also eaten when the opportunity arises. In common with gorillas (and unlike most chimpanzees) bonobos venture into swamps, pools, and streams to feed on aquatic plants. Among the plants they eat here are African flat sedge (*Pycreus vanderysti*), nimbo grass (*Panicum brevifolium*), odon (*Gambeya lacourtiana*), water plantain (*Ranalisma humile*), and the somewhat overenthusiastically named 'grains of paradise', a type of wild ginger.

Hand to mouth
Manual dexterity is important for great apes such as this young gorilla, to manipulate and position food items to find the juiciest parts and also the best angle of attack, especially for the canine, premolar, and molar cheek teeth.

Food for thought
Putting items to the lips is not always for nutrition. This bonobo absent-mindedly mouths on and gnaws at a stick, perhaps preoccupied with deciding what to do next.

Foraging strategies

Human diets may vary with the seasons – or they may not. In rich countries we have the luxury of fresh foodstuffs shipped or flown in from almost every corner of the world. Poorer nations rely much more on natural seasonal abundance. Apes in the wild organize their days and seasons around food availability, as different fruits and other items ripen in various locations around their home range. Food drives most of their large-scale movements, but other factors are also involved in their wanderings, such as courtship gatherings among orangutans, and avoiding disputes or 'trespass' among neighbouring chimpanzee troops or gorilla families.

Fruiting patterns

At any time of year the orangutan's preferred food is large, fleshy, sugar-rich fruits. However, the forests of Sumatra and especially Borneo are known for their irregular fruiting and seeding patterns. This means orangutans must adopt several ways to locate food, adjusting from 'flush' to 'famine'. They must also be able to vary their foraging patterns from one year to the next. Orangutans use several clues to find ripe fruits, including smell (especially for the powerful odour of durians) and the colour of the fruits themselves, which indicates ripeness. They also look out for the movements of other creatures, such as monkeys and birds, which might indicate a rich food source, as well as the noises of these other consumers at a feeding site. The forests orangutans inhabit have several sets of fruiting patterns. The oil-rich winged seeds of dipterocarp trees follow a 'masting' pattern, for example, with few seeds for several years, then a plentiful crop in a single season right across the forest. This pattern is linked to fluctuations in rainfall, in turn influenced by global climate cycles, especially El Niño events. The frequency of these events is thought to be affected by global warming, hence our own actions might influence orangutans' food supply. During these masting times, when food is plentiful, orangutans and other frugivores, such as the Bornean bearded pig, gorge themselves and accumulate reserves of body fat. Other trees, such as figs, may fruit two or three times annually, but these are less productive and their fruits less attractive to many orangutans, who devour them only if other foods are scarce.

Mental maps

Typically an orangutan follows a leisurely zigzag course through the forest each day, watching and listening and sniffing for foods. A productive fruiting tree may be visited for a whole day or even several days while its supplies last. The presence of other orangutans is usually tolerated (apart from fully mature males towards each other) since these great apes are not territorial. They travel widely through their individual home ranges, which overlap extensively.

Sometimes younger adults may team up and form loose travelling associations or 'parties' for the day, staying within sight of one another as they search for food. If one of them finds something exciting, the others converge. 'Partying' is most common among adolescents, who may spend up to 40 per cent of their time in these fluid associations, which are mixed-sex. Adult females spend about 10 per cent of their time in relaxed open groupings, with older males the least likely to join a 'party'. Different fruiting patterns between the Bornean and Sumatran forests have affected the amount of socializing that their respective orangutan species show. In Borneo, with its irregular, patchy food supply, orangutans tend to concentrate around trees of plenty, and so have more opportunity to interact. In Sumatra, which has steadier and more evenly spread supplies of nourishment, the orangutans tend to stay more dispersed and so interact less. Tracking observations show that orangutans build up memory maps year by year of which trees in which parts of the forest fruit when. This is part of the storehouse of information that a mother orangutan passes to her offspring over several years and the lack of the mother's knowledge presents a problem to humans tutoring orphaned young orangutans for return to the wild.

Families that eat together...

Foraging for plant foods in gorillas is, like much of their life, a leisurely family affair. The silverback (chief male) tends to lead the family from one site to the next, but if another member of the group comes upon a rich food source, such as a stand of fruiting trees or newly sprouting herbage, the others soon gather round and begin to chomp.

For western lowland gorillas the months of January to March tend to be relatively dry and fruits become scarce. So the gorillas turn to leaves – young and soft if possible – as well as twigs and other woody items, and fibrous herbs they would otherwise avoid. Important in their diet at this time are palisota (a member of the spiderwort family) and wild ginger. Compared to that of western gorillas, eastern lowland gorilla habitat has less fruit in general, so the gorillas here tend more to shoots, leaves and stems. They also climb more often than their western counterparts to get at buds and young leaves. At higher altitudes in their range they take advantage of the bamboo growing there – a food item that is a staple for the mountain gorilla. As with the other apes, the gorillas' feeding methods such as snapping, breaking and biting, act as a type of natural pruning for the plants, stimulating them to fresh growth and greater production. This is one of the many ecological roles gorillas play in maintaining the habitat. A few months after feeding in an area, a gorilla family will return to take advantage of the lush young plant parts

No hurry
The male orangutan feeds at a leisurely pace, all the time keeping senses alert for interesting events such as the long call of a rival male or the presence of a receptive female.

they have encouraged. The overall amount of food available is one factor affecting gorilla distribution, especially for the mountain gorilla. In general, for these gorillas there is enough food in their undisturbed habitat for all, all year round. As a result groups do not need to occupy and defend particular sites because they are food-rich. Instead, gorilla families wander freely and take little notice of each other. In some cases, they have almost identical home ranges, but avoid each other by using different patches of them at different times.

Social foraging

Due to their wide diets, varied habitats, often frantic social interactions, and use of defended territories, foraging among chimpanzees is a complex, fast-changing affair. Overlain on this are cultural traditions, where different groups have long-standing preferences for certain fruits and herbs which have been handed down through the generations – plus, of course, the different types of tools used to process items like hard nuts (see page 135). As chimps on the move become hungry, they use the usual cues to locate their favourite foods such as figs: smell or sight of the ripe fruits, and signs that other animals have been there feasting. Many troops split into searching subgroups known as foraging parties, and the size of these is partly affected by food availability. (Foraging parties are not the same as hunting parties, as we shall see later.) When fruits are plentiful but scattered, only in small patches, party size is larger than when fruits are more evenly spread about but less plentiful overall. This is because more numerous but smaller foraging parties can search a greater area; and if they find a limited food source, there is enough for everyone in a small group, whereas squabbles might break out in larger groups. Bonobos, like chimpanzees, travel and forage mainly on the ground, although they cover an average of just 2km (1.25 miles) daily, compared to the chimpanzees' 4km (2.5 miles). Bonobos also form foraging parties, but these are more stable in composition than those of chimpanzees: they usually consist of a family-based group of females and males plus their offspring. Most great apes obtain all the water they need from their food, especially ripe fruits. Chimpanzees drink most often, especially those occupying drier habitats in the dry season. At this time of year they may stop two or three times daily to sip or lap from a stream, a pool, a hollow in a log, or some similar place. If they cannot reach the water, they use various tools to help, such as leaf-sponges.

Ecological roles

Great apes play important roles in the ecology of their habitats by dispersing seeds in various ways. Seed dispersal is vital for young plants if they are to get away from the shade cast by their established parents and find the space and light to grow. Most apes are fussy, messy eaters and discard a proportion of the fruits, nuts, and similar items they collect as they select the tastiest. They sometimes do this on the move, travelling dozens or even hundreds of metres while carrying their booty, to find a secluded spot where they can check the meal and consume or reject it. This is one way in which the seeds of the plants they feed on are spread.

Caring sharing
In line with their generally non-confrontational lifestyle, adult female bonobos are known to share food with youngsters – including those who are not their own offspring.

Feeding routine: orangutan

For most days of most years, excluding breeding activities, the orangutan's daily life revolves around food and feeding. The foraging strategies of the great ape species are described on previous pages. How does this affect their daily routine and what they decide to do when, day after day?

Like all apes, orangutans are primarily visual creatures engaged in few activities during the hours of darkness. Observations show that, in general, they night-sleep for two-fifths to one-half of every 24 hours. Added to this is a restful doze, usually after the hard work of the morning feed. In this routine, the red ape is very similar to the gorillas. Dozing in the night nest high above the forest floor, the red ape is safe from almost all natural hazards. Most orangutans sleep on their sides, using their arms to support and cushion the head. However heavy adult males are known to spend more time on the ground and even sleep there in the manner of male gorillas. With no light switch to flick, their lives are ruled by the rising and setting of the sun in a way we can only dimly imagine, even when we fall victim to power blackouts.

In the early part of the morning on a typical day, an orangutan listens to the dawn chorus – often more of a dawn cacophony – with the calls and songs of many creatures. Especially

6.20AM
Dawn in the equatorial rainforest. Already awake in the nest, listening to calls of birds, monkeys, gibbons, and especially the long call of a nearby male orangutan.

6.50AM
Grooming, defecation. Sniff the air to determine smells of ripe fruits, especially durian, in the vicinity

7.20AM
Continue to doze but remain aware of scents and sights, especially the movements of other animals

9.05AM
Slowly move through the mid-canopy to a nearby durian tree, begin to feed

9.30AM
A neighbouring female arrives at the same tree, but since the fruits are plentiful, there is no aggression and both feed in different parts of the tree

9.55AM
Pause to rest and listen, grooming. Search for bromeliad flowers with ripe shoots and buds which have also trapped fresh water in the bowl shape of the leaves

11.10AM
Further feeding on durian, but the arrival of another female means rising possibility of some aggressive behaviour

6 am 7 8 9 10 11

noteworthy is the penetrating 'long call' of a nearby male, which could signify his intent to locate a mate. Stirring from the nest after daybreak, although perhaps up to a couple of hours later in the mid-morning, the priority is breakfast after a long night without sustenance. However any kind of agitation, and especially the alarm calls of birds, may cause early alertness, with ears straining and eyes roving around the canopy to the forest floor below. The reasons are numerous – natural predators such as a tiger on the prowl, a clouded leopard or perhaps a large python sneaking through the branches. Sadly, in modern times a major cause of disturbance is approaching human poachers on the lookout for any profitable game in the area. Then the orangutan's priority is to steal away from the trouble as quietly as possible. For such bulky tree-dwellers, orangutans can move with amazing stealth and silence, hardly shaking a leaf as they ease their way and disappear into the green blizzard of the canopy.

Party time

The orangutan's social life, especially when compared to chimpanzees, bonobos and ourselves, is not what might be termed hectic. Typical daily activities during waking hours show how the orangutan spends much less time socializing or disputing with its own kind (intra-species contact) compared to chimpanzees. But the red apes are not aloof from an occasional bout of 'partying' when a few juveniles or younger adults gather in a loose aggregation for a time, varying from a day to a week or more (see page 151). The usually stimulus is plenty of fruiting trees concentrated into one small location. Natural selection has come up with the solution that tolerating the presence of others is the best strategy while there is lots of food to share around.

daily activities: chimpanzee

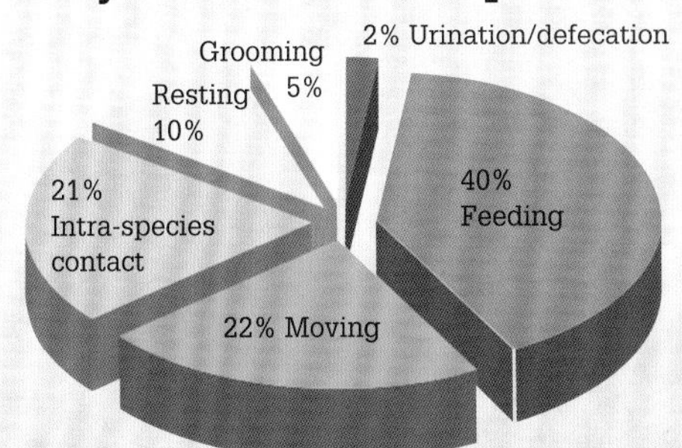

daily activities: orangutan

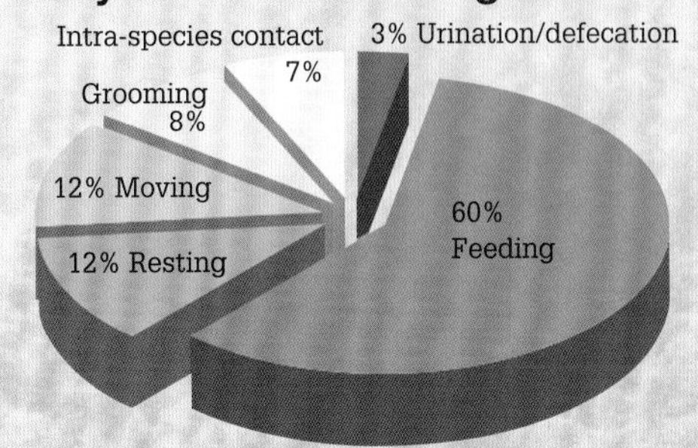

11.45AM
Noise and calls indicate an approaching male orangutan. Rapid movement to a nearby non-fruiting tree as not ready to mate

11.55AM
Pause for rest in noonday heat, keeping senses alert for dangers such as leopard, tiger and, in the branches, python

12 noon

12.35PM
Move to mid-lower canopy to avoid harsh sun. Midday grooming session

1pm

1.15PM
Sun obscured by clouds. Move to upper canopy to keep cool in developing breeze

2

2.30PM
Careful journey through the mid-canopy to where movements and sounds of hornbills and monkeys indicate a fruiting strangler fig

2.40PM
Feeding on strangler fig

3

3.35PM
Come across tree termite nest, break open to eat the grubs, then gnaw soft bark for sap and juices

3.55PM
Resume fig eating

4

4.30PM
Search begins of 3–4 trees to find suitable night nest site

4.50PM
Tropical storm postpones search. Rest under a branch with thick foliage for some protection

5

5.20PM
Night-nest tree selected; branches bent over to make platform. Evening grooming session. Listen to dusk chorus to identify any orangutan calls

6.05PM
Darkness. Settle to rest

6

Meat-eating

Tapas, meze, canapés, antipasti, hors d'oeuvres, smorgasbord – tasty bits and pieces of snacks, often containing varied meats, feature on many of our national menus. So it is with wild apes. In between bouts of munching fruits and vegetation, they take any opportunity to feed on all kinds of varied small animal foodstuffs, from tiny insects to lizards and rodents such as mice.

Orangutans, once thought to be the most exclusively herbivorous of the great apes, consume a long list of miscellaneous small creatures. Insects and other small arthropods predominate, including crickets, leafhoppers, termites, ants, spiders and their webs, various types of worms – as well as the eggs and grubs or larvae of these creatures. In short, they eat almost any small invertebrates they come across.

As with the other great apes, orangutans sometimes swallow these small creatures accidentally – a caterpillar hiding in a curled-up leaf, for example, insect eggs laid on a stem or under bark, or beetle grubs in decaying wood. Even when they deliberately pick up the food with their fingers or mobile lips it is usually opportunistic, because they happen to find such a food item as they feed on their plant staples.

Bornean orangutans are known to devour bird and reptile eggs, baby birds from the nest, and small mammals such as mice and squirrels – again, especially the young from the nests. Sumatran orangutans take advantage of these items less often, although there are instances of adult females eating slow lorises, which are smaller and more primitive but nevertheless their primate cousins.

Wild honey is much enjoyed by orangutans, but they are less brave than chimpanzees when raiding bees' nests for it and soon retreat. Mushrooms and other fungi also feature in the diet of orangutans, as well as those of other apes.

Chimpanzee fast food

Depending on the area and season, the diet of some chimpanzees can be up to one-tenth miscellaneous small animal fare. This far outweighs the amount of flesh they consume from larger creatures caught by hunting. Termites are a perennial favourite, and the regular subject of tool-use. Termite adults, grubs, and eggs are all consumed. Weaver ants are another common prey, licked off leaves and twigs, and quickly swallowed before they can use their sharp mouthparts or stinging chemical defences to protect themselves. Social insects such as wasps and bees also occur on the menu. Once a nest has been found, dozens can be consumed with relatively little effort. This makes them 'cost effective' in the energy equation – energy contained within a food compared to energy expended obtaining it.

As well as termites, ants, and other social insects, chimpanzees greatly prize wild honey, as do bonobos. The usual source for honey is the nests of stingless bees. In some areas chimpanzees and bonobos take to the trees to find the nests, harvest the honey, and munch up bee larvae, pupae, and adults – in fact the entire nest is a meal in itself, rich in sugars, proteins, and other nutrients. In other regions chimpanzees and bonobos prefer to search for bee nests on the ground and dig them out either by hand or with digging sticks. Honey is so desirable that some chimpanzees endure the stings of other bees to secure this delicacy.

A mine of information

Faecal analysis may sound unappetizing, but it's a very important method of discovering what apes and other creatures have eaten. It is useful not only to know more about their ecology and requirements in the wild, but also to arrange healthy diets for captive apes. Faecal analysis reveals that, in general, chimpanzees eat termites, ants, or both about once every two or three days throughout the year. Dozens

of other insects feature in the chimpanzee bill of fare, from grasshoppers and beetles to cockroaches and caterpillars.

Both chimpanzees and bonobos are recorded as scavenging on a variety of dead or almost-dead small animals including bats, lizards, and even, in one bonobo case, a duiker (a type of small antelope). Bird eggs and nestlings are also snacked upon when the opportunity arises.

Gorilla fare

As with other aspects of their diet, the western and eastern lowland gorillas show some marked differences in their small animal intake. Western lowland gorillas are known to take more than 25 different types of insects and similar invertebrates. As with chimpanzees, termites and ants top the list. One study showed that some western lowland gorillas eat at least a few termites every day. Caterpillars, grubs, and other assorted items made up the bulk of the rest of the

Chance finds
An orangutan swipes some termites found on a stick. Almost any small, juicy bug, grub, egg, or creepy-crawly that presents itself may be consumed. There's no sense in letting good food go to waste.

Tiny tid-bits
Grooming is partly social, partly hygiene-based – and occasionally yields micro-snacks of sustenance in the form of lice, fleas, and other skin parasites.

invertebrates eaten, often swallowed by chance along with the main plant food. Faecal analysis of gorilla dung has detected insect remains in up to one-third of droppings.

As with plant foods, there are signs of cultural differences between western lowland gorilla populations as regards their small animal fare. In one region the gorillas prefer weaver or green ants, while a few hundred kilometres away they go for termites – even though both insects occur at both locations. The implication is that younger members of a gorilla group watch the older ones and imitate their actions, and in this way traditions and food preferences are passed on.

Birds' eggs and nestlings, small reptiles, such as lizards, and small mammals, such as mice, are all eaten by gorillas. However, these are exceptional events, and overall, the average western lowland gorilla's intake of animal matter is much less than that of a typical chimpanzee.

Eastern lowland gorillas have a similar pattern of small animal snacking to their western counterparts, but their overall intake is less. Again termites and ants predominate, while other animal items are extremely rare, especially small vertebrates and their eggs.

Of all the great apes, mountain gorillas probably consume the least small animal matter – some studies put the amount at only 0.1 per cent of total intake. (On average, a mountain gorilla consumes 25kg (55lb) of food a day, so that equates to just 25g (1oz) of animal matter in a 24-hour period.) Ants, termites, grubs, and worms all feature on the menu, and individuals have been seen breaking into ant and termite nests. Larger items such as birds' eggs and young, as well as wild bee honey, are available but mountain gorillas seem to be oblivious to them.

Dirt as food

All wild great apes have been seen to eat something that few humans would relish – soil. This practice is known as geophagy, and there are several theories about the reasons why it happens among wild apes:

- Soil provides mineral supplements to augment the largely herbivorous diet with tiny amounts of health-giving substances such as iron, manganese, magnesium, and zinc.
- It acts as a medicinal aid by neutralizing mild toxins in foods that could cause what we would call an 'upset stomach'. Orangutans eat soil rich in kaolin (china clay) which would absorb chemicals such as tannins that plants produce to deter browsers.
- Some soils contain a high proportion of animal matter such as insect and worm eggs, tiny larvae, and other items. Eating such soil would provide wild apes with protein and other nutrients. Like other animals, from elephants to rats, orangutans are also known to visit 'salt licks' where the soft mineral-rich rocks provide substances such as calcium, sodium, and potassium, which, ordinarily, would be lacking from their diet. Among some populations, visiting salt licks may even replace geophagy completely.

Coprophagy

If eating soil seems distasteful to humans, coprophagy is positively repulsive. This process is the consumption of faeces or dung, either one's own or that of another type of animal. All wild great apes are known to indulge in coprophagy. Wild gorillas and chimpanzees have been seen eating the faeces of various animals, including themselves and their fellow troop members. Faeces consumed are usually hours old and dried, but not always. As with eating soil, the suggested reasons for coprophagy revolve mainly around obtaining important mineral nutrients. In one study group of chimpanzees, coprophagy was linked to feeding on velvet tamarind fruits. It was found that after one passage through the chimpanzee gut, the seed casings were softened and damaged. If the seeds were well chewed before being eaten the second time around, the chimp's digestive system was then more likely to break them apart and extract the rich nutrients they contained. (Coprophagy in captive apes is a different matter from that seen in the wild. Here it is usually assigned to problems such as stress or a diet that is inadequate in essential nutrients.)

Natural need
This chimpanzee instinctively tops up its dietary intake of minerals by hand-cupping loose, sandy, salt-rich soil from a regularly utilized source.

Balanced meal
A young mountain gorilla munches on a root and in the process consumes any worms or grubs that may be living in it, as well as mineral-rich soil particles coating it.

On the hunt

In the early 1960s at Gombe, Tanzania, Jane Goodall first reported the existence of cooperative hunting among chimpanzees. She had witnessed these wild apes working together to capture large animals and consume their flesh. Her reports made headline news. People were fascinated by the implications for our own evolution from distant ancestors who were mainly herbivorous through the hunter-gatherer phase to today's mass farming of livestock. The fact that chimpanzees were hunters helped to explain our own very carnivorous diets, as well as giving an insight into the origins of our tendencies to form hunting parties, to pursue large prey, and to kill.

The victims

Among the wild great apes, only chimpanzees are known to organize communal hunts for larger animals. The creatures they target are nearly always mammals. Some of these animals are hunted on the ground and others up in the trees.

Since the first reports from Gombe, chimpanzee predatory behaviour has been recorded at several other sites across the species' range, including Taï in Côte d'Ivorie, Kibale in Uganda and Mahale in Tanzania. The total list of victims stretches to more than 30 different species and includes several types of monkeys, such as colobus, guenons, vervets and young baboons, as well as bushpigs, small antelopes known as duikers, larger antelopes called bushbucks, flying squirrels and tree pangolins (armoured mammals that look rather like a cross between an anteater and an armadillo).

The numbers of hunts, who takes part, and their success rates vary across these regions, but in ways which are not yet well understood. Seasonal changes and the availability of other foods certainly play a part, yet in Gombe the main hunting peak is in the dry season, while at Mahale it is at the start of the rains. Chimpanzee community size and its make-up in terms of males, females, and young also appear to be involved. A predominance of adult and adolescent males in a group usually means that hunts happen more often. Large groups are also more likely to hunt than small ones, as the former contain more potential hunters. Individual personalities also play a part: some large males instigate hunts often, while others rarely do.

In 'binge' years hunting becomes far more popular than usual, but again, exactly why this happens is not clear. In a binge year a typical community of 50-100 chimpanzees will kill more than 150 victims, representing more than 600kg (1320lb) of dead meat. Yet for the same group in other years, and for some groups almost always, the annual toll is fewer than 20 kills.

Anatomy of a hunt

The most popular prey for chimpanzees at Gombe are red colobus monkeys (*Procolobus rufomitratus*). They make up four-fifths of all quarry, and of those red colobus caught, about three-quarters are youngsters. The hunting party varies from a few individuals, or even just one, to more than 30. On average, 90 per cent of the hunting parties are male chimpanzees, either full grown or adolescents. In general, the more individuals there are in a hunting party, the more successful it is likely to be, with success rates exceeding 90 per cent for the biggest parties. Sometimes the hunters actively target and chase prey, while in other cases they gather around en masse after happening on a victim by chance.

Division of labour during the hunt itself has been most studied among the Taï chimpanzees. Once the prey is targeted, it appears that some individual chimpanzees act as 'drivers' to move it in the desired direction, while others are visible 'blockers' cutting off potential escape routes. Along with the drivers and blockers are hidden 'ambushers' and 'captors' who initiate the kill. The complex coordination of hunting roles takes time to learn. Taï chimpanzees start at about 10 years of age, but are not usually proficient for another 10–20 years.

Sharing the kill

What happens to the hapless colobus monkey or other victim? Usually it is torn apart by a chief male while it is still alive and shrieking, as excited chimpanzees gather around for a share, agitating and begging. Taï chimpanzees are even known to bite and break the prey's bones in order to get at the nutritious marrow within.

Like the reasons for hunting, sharing the spoils is a complex business. Generally the dominant males who were most important in the hunt split most of the reward between themselves. But meat from these cooperative kills often represents far more than just food. It has social, political, and sexual overtones. In one group an alpha male chimpanzee was known to share meat with his allies in the hunting party, presumably to reinforce his status and dominance, while keeping it from his rivals. Another study of a different group showed that hunts were more likely to occur when there were more sexually receptive females around; also males were more likely to share food with these females if they begged; and the females themselves were more likely to produce babies that survived. In this way hunting prowess could feed through to reproductive success, for male chimpanzees at least.

Courting the chief
Famous Frodo (born 1976), alpha male at Tanzania's Gombe Stream site from 1997 to 2002, takes charge of a bushbuck fawn carcass. Other chimps gather around, hoping both to share the spoils and to reinforce their social standing with the boss.

Communication Skills

Self-imposed solitude can be good for the human spirit. But for most of us, there is a limit to how long we can go without human contact. Likewise a solitary chimpanzee, bonobo, or gorilla is an unusual sight in the wild. These great apes are creatures whose social lives are built on endless communication with their fellows, using all sensory modes from gentle tactile caressing to pungent odours, intimidating visual displays, and blood-curdling screams. Compared to other great apes, orangutans are loners, especially adult males. Yet even they use various modes of communication, to locate breeding partners and establish territories, home ranges, and hierarchies.

Why communicate?

Animal communication is a vast subject, but one that is basically divided into two distinct parts, known to zoologists as interspecific communication and intraspecific communication. Put simply, interspecific communication is the communication that occurs between species. Examples include rearing up or the baring of teeth to deter predators.

Intraspecific communication occurs between members of the same species. Most animals need some form of interspecific communication, usually for reasons of simple survival. Messages to members of other species tend to be more obvious and less subtle than those used for dealing with conspecifics (members of the same species). Interspecific communication is more of a blunderbuss approach and often involves behaviour that occurs widely across many animal groups. The hiss, for instance, warns that an animal is ready to defend itself. It is a sound employed by creatures as varied as cockroaches, bees, spiders, toads, snakes, lizards, skunks, and cats.

Seeing off predators

Great apes do not employ defensive hissing when faced with a predator. They tend instead either to engage in all-action confrontation or make a sharp exit. One of the chimpanzee's few natural enemies is the leopard, which sometimes targets infant chimps as prey. A series of ingenious experiments carried out in the wild used a dummy leopard with a moving head and tail, and a baby chimp doll in its paws for good measure, to reveal how groups of wild chimpanzees communicated their displeasure at this threat. Their general tactic was to rush around and jump in an agitated manner, while making shrieking or screaming 'wraaa' alarm calls, to show that they were ready to attack. The chimpanzees then picked up sticks or broke small boughs from bushes and threw these at the enemy. Nearby monkeys and other animals were alerted by the commotion and fled. If the leopard had been alive it would almost certainly have done the same, dropping its victim to hasten its escape. Gorillas have very few natural predators, again apart from leopards, and perhaps crocodiles in lowland areas. When gorillas need to communicate their intent to attack in defence of themselves or their young, they tend to use modified versions of the rushing, vegetation-thrashing and chest-beating displays that are usually intended to drive away intruders of their own kind (see page 192–3).

Reasons for communication

Intraspecific communications in great apes, as in many other animals, can be grouped according to motives and aims. Most noticeable to us are the aggressive or agonistic messages given, including intimidation or threats over food, territory or mating rights. The aggressor's aim is to appear as big, fierce and forceful as possible. These sounds and actions often look frighteningly hostile, but in the vast majority of cases they allow apes to quickly sort out matters without actually resorting to physical conflict, with all of its associated risks. The aggression is highly stylized, or 'ritualized', and clearly understood by conspecifics. In ritualization, a behaviour becomes separated from its original function, modified, and stereotyped to fulfil a new signalling function in communication. Ritualization is not confined to behaviour concerning conflict. For example, a chimpanzee intending to play may slap the potential playmate gently; this was perhaps an aggressive action originally, but it has now been modified into an attention-getting message saying 'Let's mess about!'. In some cases simply raising the arm is enough for the playmate to understand the situation and play to be initiated (see page 196). Another group of intraspecific communication deals with courtship and preparing to mate, and here again ritualization plays an important role. If sex is to result in reproduction, it is as well to make sure your desired partner is of the same species as yourself, of the opposite sex, and as healthy and fit as possible. Following successful mating are the other very special communications that occur between babies and their mothers, and between growing infants and other members of their groups.

Food is central to ape life, and the social great apes may call their offspring or special friends to a food source, or solicit food from another group member for various reasons. There are also alarm and warning communications made for the benefit of the rest of the group, telling them trouble is brewing.

Communication and language

Over the years there has been much debate over whether non-human great apes have true language. Much of the argument centres on the definition of language: some say true language requires formal structures such as grammar, syntax, and combinations of words with novel and specific meanings (or phrases). Studies of captive great apes reveal fascinating insights. While in their natural habitats the great apes communicate in amazingly complex, extensive, and subtle ways, without computer screens, keyboards, or lexigram symbols.

Let's pretend
(Previous page) Juvenile male gorillas play-fight. The open mouth, upright stance, raised arms, and vocalizations will develop over the years into one of nature's most fearsome displays – the mature silverback's threatening charge.

Pulling faces
Juvenile and adult bonobos encourage each other to make strange facial expressions. They even pull faces when alone.

Watch with mother
The orangutan youngster learns some of its communication skills from its mother, although an older sibling may help.

Visual signals

With a slightly raised eyebrow or the merest wave of a finger, humans can communicate all kinds of information by sight. So can the other great apes. Many decades of research have revealed a vast range of visual communications within the ape species – body language, facial expressions, physical gestures with arms and legs, speed of movement, and how exaggerated or subdued this is, and much more besides. However, unwary human observers have encountered many pitfalls along the way.

The chief accusation levelled against observers has been that of anthropomorphism – attributing human characteristics or traits to non-human animals. Confusing non-human great ape behaviour with our own is easier than one might imagine. For example, a chimpanzee with mouth pulled wide and teeth visible might appear to us to be smiling or grinning – something we associate with pleasure and enjoyment. However, a chimpanzee making this face is much more likely to be frightened. The chimp 'smile' is actually a fear grimace – a face that signifies an individual is afraid or agitated in some other way. The wider we grin, the more amused we are; the more marked a chimpanzee's 'grin', the more intense is its fearful state.

Messages in context

In the earliest studies of ape communication, researchers tended to isolate one visual aspect, such as mouth shape, and observe its meaning by the reactions of other group members. What has since become clear is that most aspects of communication are 'multimodal' and 'contextual'. In other words, a single message is sent in several different modes, which can include body posture, facial expression, vocalization, and movement, and its meaning depends on the context or given situation at the time. An example is the bonobo 'grin'. In contrast to the marked differences in vocal communication between chimpanzees and bonobos, the facial expressions of these two species show many similarities. Lips pulled back into a grin can show fear in both species – but this is not always the case. Bonobos may put on a grin-type expression to show satisfaction, for example, when having sex, examining and approving a strange object, or completing a task such as building a night nest. The meaning of the bonobo grin (unlike that of the chimpanzee) depends on the context in which it is given.

In one detailed study comparing bonobos and chimpanzees, hundreds of hours of observations were analyzed for communications involving combined facial expressions, vocal sounds, and arm and hand gestures. Only the communications that produced responses in other group members were considered, since these apes often 'pull faces' when on their own, seemingly for their personal fun and amusement. The results showed that in both species, all three modes of communication varied according to the social context, such as food sharing, a dominance dispute, or wanting to play. However, gestures varied more than did faces or sounds. Also when combining these three modes, bonobos used gestures and faces together about as often as they combined gestures and sounds. Chimpanzees, however, favoured gestures with sounds over gestures with facial expressions. Another finding was that bonobos produced more responses in their fellow group members when using all three modes of communication together than chimpanzees did. This all goes to show that we still have a lot to learn about ape communications that are multimodal, combining vision, sound, and perhaps touch and smell.

Making faces

Of all the non-human great apes, chimpanzees and bonobos show the greatest range and subtlety of visual communication. In most situations a long stare from a chimpanzee means some kind of tension or threat, especially if combined with erect hair. A lip pout, as though puckering up for a kiss, is used to ask or beg for food. Letting the lower lip sag or hang indicates relaxation and calm. A lip-smack greeting is used as a sign of friendship and submission, often in combination with grooming. The play face with mouth slightly open and an otherwise relaxed demeanour indicates that all is well.

These facial expressions are often combined with visual signals from other body parts. An angry chimpanzee stands up, waddles, or swaggers, and waves its arms about, perhaps even throwing stones or branches. This display is often combined with screaming noises and

Looking big
Standing tall with limbs held out and hair erect, in order to appear larger and more daunting, is a common mammalian behaviour, here demonstrated by an aggressive male chimpanzee.

Weaponry
The chimpanzee's tool-using inventiveness extends to weapons; this agitated individual commandeers a stick to threaten an intruder.

Interested
The chimpanzee's expression shows attention and concentration, when studying a scene or situation, making mental notes and ready to react with increased arousal or disinterest.

Aroused
A state of arousal or excitement, with lips pouting or pursed, an intense gaze, and the body tense and poised, able to spring into sudden action.

rushing to and fro. A male chimpanzee in a hierarchy dispute with another male will stamp, slap his hands on his body or on nearby objects, and drag branches and thrash vegetation. This is all designed to make him look as threatening as possible, to intimidate his rival in a ritualized way without resorting to a physical fight.

Bonobos vary their basic facial expressions for certain types of social interactions, for example, sorting out their position in the group hierarchy. An individual confronted by a more dominant group member may put on an open-mouthed grin, pulling back the corners of its mouth partially to reveal the teeth, in order to show fear and submission. Meanwhile, the dominant bonobo may open its mouth but keep its teeth covered by the lips and stare hard. This open-mouth threat expression is used to keep the subordinate receiver in its place. Eyes wide and mouth open but with no teeth showing is the play face, as in chimpanzees.

A useful tool

In the 1970s scientists devised Human FACS, the Facial Action Coding System. This is a system for dividing any facial expression into about 30 building-block parts, known as action units, each of which is based on certain muscle contractions. Action units are quantified using biometrics (the science and technology of measuring and analyzing biological data) in order to describe the expressions objectively. It is then up to an interpreter to decide what the particular expression means in any given situation or context. Human FACS was revised and updated in the early 2000s, and it is now being expanded and adapted for use with chimpanzees. Early results from Chimp FACS have shown that chimpanzee facial expressions are much more subtle and delicate than we previously imagined. Their meaning can be altered depending on the detailed positions of the brows, eyes, lips, jaw, and other facial features. For example, bared teeth can be incorporated into several different expressions, depending on the positions of the lips, how far the mouth is open and how high the brows rise. Each of these expressions can then be used to express different meanings in different contexts. Researchers on this project include chimpanzees themselves, who are asked to use their 'insider knowledge' and match three-dimensional animated expressions shown in images for content and intent. Chimp FACS data has been cross-checked against decades of human observations and interpretations, and the two have been found to correlate well. As an objective tool standardized for use around the world, Chimp FACS could open a new era in understanding what chimpanzee faces mean, and perhaps tell us more about human expressions and emotional communication too. It would be extremely interesting to know whether our own facial communications, and those chimpanzees and bonobos, have the same roots – in our common ancestors from more than five million years ago. Chimp FACS may help us to discover whether or not this is the case.

Gorilla social life is generally much less hectic and complicated than that of chimpanzees and bonobos. But these greatest of the apes still have their own wide array of facial expressions, including a play face with widened eyes, a yawn that indicates stress, and an aggressive stare while remaining still, which the silverback uses to great effect when quashing subordination in the ranks. One of the most evocative expressions, as far as humans are concerned, is the

Playful
Largely relaxed, with the mouth partially open, the upper lip covering the upper teeth and the lower lip hanging slightly to expose the lower teeth.

Fear
Mouth wide open, lips pulled tight to bare both upper and lower teeth; sometimes called the 'fear grin' and compared to the human's 'nervous smile'.

young gorilla's slightly protruding tongue, seen during concentration or when an individual is feeling uncertain. This is mirrored in our own species and seen when children are thinking hard, or absorbed in some new and challenging task.

The show of hands

As well as faces, hands play a crucial role in great ape visual communication. Interestingly, some scientists suggest that chimpanzee and bonobo hand gestures could hold the key to how our own spoken language evolved. The idea is that hand gestures are 'prelinguistic' versions of phrases, used to express thoughts and send messages, from a time before we could speak. If you have trouble grasping this concept, think of the ease with which people 'talk' with their hands today. Communication through hand gestures takes place not only in organized sign-language systems but also casually, such as when shaking a fist to express anger or holding out an open hand palm-upwards to request an object.

Studies of bonobos and chimpanzees reveal that the use and meaning of hand and arm gestures is extremely variable, compared to facial expressions and vocal sounds, which tend to be more standardized and stereotyped. These apes demonstrate more than 30 basic arm or hand positions and movements, and continually modify them, not only according to the situation, but also for different individual recipients, paying greater degrees of attention to some than to others. Gestures also become personalized and get copied by others in the group, especially among bonobos, as a form of cultural transmission. The many ways in which we, chimpanzees and bonobos use and modify gestures could mean their basis dates back to our shared ancestor.

Fear not fun
A classic case of how we can come unstuck reading human appearance and behaviour into other animals is the chimpanzee's 'grin'. It is actually an expression of agitation and fear.

Messages through sound

Often described as one of nature's most distinctive and eardrum-penetrating sounds is the famous 'long call' of the fully mature male orangutan. Males call about three to four times daily and more often where their population density is higher. The call commences softly with a series of low resonating rumbles or bubbly grumbles; it then builds with tension, becoming louder and more forceful, into a full-throated, gravelly roar or bellow that reverberates through the forest for a kilometre or more.

The call is enhanced by the male's hollow air sacs (laryngeal chambers), which act as resonators in the same manner as the pipes of a church organ. His cheek pads or flanges may also help by directing the sound like a megaphone so that it projects forwards. As he roars, the male's hair stands erect. He may also shake the branches around him for good measure or even pull down a partly broken branch to cause a crash that adds to the effect. Two minutes or more after starting the show, gradually the roar begins to fade in power into a series of softer, more yielding 'breathy' sighs and moans. Finally it peters away and the jungle can relax and resume its normal business.

A proclamation of intent

Why does the male orangutan make such a tremendous communicating sound? Put simply, the long call announces his presence as the local boss to other males who might think of muscling in on his territory. It helps mature males to space themselves suitably through the neighbourhood, and in doing so reduces physical encounters. Since each male's territory includes local females, the call also warns other males that these are spoken for. At the same time, it reminds the females that the male is still around and ready to father their next offspring. The long call is often given when a male arrives at a new site in his territory, and after mating with a female. Sometimes one male's call triggers his neighbour to a similar proclamation, so that the two exchange information about their relative locations.

In the 'sonic kingdom' of the dominant male orangutan, there may well be junior males within earshot who could be direct rivals but are not. Unlike dominant males they lack cheek pads and many other secondary sexual characteristics. The reason for this is thought to be that the long calls of senior males keep the junior males' stress hormone levels elevated, which prevents them from developing further. This phenomenon has been dubbed 'acoustic castration'.

Orangutans live high in dense forest, and the meanings of their other calls are difficult to ascertain. They emit a wide variety of vocalizations including squeaks, barks, screams, grunts, rumbles, and burps, as well as non-vocal noises like lip-sucking, teeth-grinding when frustrated, lip-smacking when in a group, and bough-slapping. One sound, the 'kiss-squeak', is executed by sharply inhaling air through pursed lips and conveys agitation, fear, or excitement. There are also various types of screams that signify worry, anxiety, threat, and fear. Other vocalizations are associated with distress, especially in infants and juveniles, and with contentment, play, and mating.

An enigmatic chorus

Unlike orangutans, the sounds gorillas produce are often relatively subdued and enigmatic. They include a wide range of grunts, grumbles, coughs, snorts, belches, hiccups, hoots, barks, screams, and roars, which are distinct to each individual. Presumably some of the quieter, everyday sounds help the group to know each member's whereabouts as they feed, rest, or travel in thick forest. But matching particular sounds to particular behaviours or listeners' responses is tricky. In one series of observations on mountain gorillas, the adults vocalized every five to ten minutes on average. More than half of the sounds were parts of exchanges with other group members, especially those within about 5 metres (16 feet). In general the dominant adult males were noisiest, followed by the adult females, then the subordinates and youngsters. Adult sounds were more grunting and guttural, while the infants made lighter, more melodious sounds. But many calls that were recorded went unanswered or apparently unacknowledged, so whether they serve to communicate to others is unclear.

Among the mountain gorilla's better-known sounds are a hoot-like call that carries well through the forest, and various grunts and pig-like snorts, which are used to maintain contact while out of sight or moving around during foraging. Double grunts are used in various ways; a version with the second grunt in a lower pitch may occur after a quiet period, and it can elicit a reply, often another double grunt but with the second section in higher pitch. However, similar grunts seem to be given in various situations: whether they are context-specific, having different meanings depending on the circumstances, is

Weighing up the situation
(Prevous page) A young orangutan approaches an unfamiliar elder slowly and warily. Each sizes up the other and notes postures and movements, to assess whether there is any potential threat.

Warning sounds
The silverback mountain gorilla beats his chest, warning that he is ready and more than able to defend his group.

Chimp talk
Chimpanzees have a wide range of calls based on in-out panting breaths. These grade from a slow, soft 'hoo' through grunts and barks, to ear-piercing screams – sometimes all in one unbroken sequence.

not clear. As is the case among the other great apes, young gorillas whimper for attention from their mother or when in distress, and make noises uncannily like breathy human laughter and chuckling as they play. Group members may also produce a curious humming or purring when relaxed or feeding contentedly. Louder barks signify some kind of rising tension or excitement, especially from blackback males who detect danger or intend to challenge one another. Bigger threats cause louder bellows and screams, culminating in the full roaring charge of the silverback, described on page 192.

Chest beating and thumping are common non-vocal sounds used by gorillas, especially the males. They are generally used to intimidate outsiders, intruders, or rivals for dominance.

The chimpanzee repertoire

Compared to orangutans and gorillas, chimpanzees have busier, more complex social lives, and a greater variety of communications to match. Researchers have produced various lists of vocal calls and their meanings, from a dozen to 30 or more, but admit that it is difficult to precisely define them. Some calls grade or segue together, while others are intermediate in form, sometimes sounding nearer to one of the more 'standard' calls and sometimes more like another. For instance, as a younger male nears an older, dominant male, he may pant-grunt, then increase the tempo and volume into pant-barks, and even into pant-screams as he becomes more nervous and fearful.

One of the loudest chimpanzee calls, and the one many humans associate with the species, is the pant-hoot. It starts with low-pitched hoots that merge into faster in-out panting, and perhaps into a loud crescendo. Each individual has his or her own pant-hoot versions, which are recognizable to humans as well as to other chimpanzees. Pant-hoots are made most commonly by senior males, often to keep in touch with their allies, and carry well through the forest. Females also pant-hoot, sometimes together in a chorus. This call can be made while on the move, in response to other distant calls, and when encountering a good food supply or other group members. Pant-barks and pant-grunts tend to show submission when being approached by a more senior group member.

Screams and whimpers

A screeching or screaming chimpanzee is usually being threatened, pursued or physically attacked. Young chimpanzees whimper for their mothers, and like chimpanzee screams, these vocalizations can sound unsettlingly human. A loud 'wraa' often indicates fear, 'hoo' curiosity or puzzlement, and panting or lip-smacking enjoyment. There are also food-related calls such as the food-grunt or the 'aaa', which shows food satisfaction.

These are general chimpanzee call categories, and there is much more for us still to learn. Analysis of calls at Taï Forest, Côte d'Ivoire shows how complex chimpanzee vocalizations may be. For example, bark messages are used in different contexts, such as when hunting monkeys or warning about different dangers. Combinations of different calls can also be supplemented by non-vocal sounds like drumming on tree buttress roots. More than 80 calls and combinations have been noted at Taï. In some combinations, single calls from certain contexts were added to make a combined call with different information for a new context. In addition, vocalizations may well be accompanied by visual messages such as facial expressions, body postures, and hand gestures. It is a complicated business being a noisy chimpanzee and an even more complicated one being a researcher trying to find out what all their vocalizations and visual gestures mean. Adding to the information gleaned from Taï Forest, studies at Gombe and Mahale, both in Tanzania, show that chimpanzees from the two locations have slightly different pant-hoots, which could be the equivalent of local dialects or accents.

As already briefly mentioned, non-vocal sound messages are also used by chimpanzees. These may be conveyed by sounds such as drumming or slapping on tree parts or other objects. In leaf-clipping, a chimp gathers some stiff leaves and nibbles or bites small pieces from them with a ripping or tearing sound, but without swallowing. At Mahale, this can signify readiness to mate, but at Taï it is carried out by males to assert their seniority.

The noisiest ape

Chimp and bonobo vocalizations have many similarities, but also important differences. Overall, bonobos are probably the noisiest wild apes. They communicate over long distances by 'high-hoots', which vary in pitch but tend to swell in volume and can sound like dog yaps. The hoots are often replied to very promptly, so that a fast exchange builds up and telescopes in time to produce an eerie multi-echo effect. These messages can ring out at almost any time, but are especially common when individuals are establishing a feeding area in the morning or a nesting site later in the day. They probably serve for group cohesion, telling each member where the others are.

The bonobo's submissive sound is a 'koo-koo', produced by individuals when approaching a more dominant group member. If two individuals challenge one another, there is a rapid burst of varied calls mixing threat with reconciliation. Sometimes the tone and volume of each protagonist's calls becomes gradually more similar to the other, a phenomenon known as vocal matching.

In addition to their calls bonobos have many forms of non-vocal auditory communication such as tree-slapping, shaking bushes, chest-beating, and episodes of short but fast drumming with both hands and feet on tree buttresses. In fact, the only time that bonobos are really quiet (apart from when they are asleep) is when travelling on the ground to a new site, when the small parties maintain a form of 'radio silence'.

Call of the forest
Each individual orangutan's vocal repertoire is familiar to those within earshot; the dense canopy precludes visual recognition.

Apeing each other
Two chimpanzees encourage each other as they vocalize to show their disapproval of a trespasser.

Touch, scent, and smell

Hands clasp tenderly, fingers caress the cheek gently, lips approach eagerly … the romantic novel's stock-in-trade is communication by touch, at least for its human characters. We are a socially tactile species, both by instinct and learning. The other great apes also indulge in much tactile communication with a wide range of physical contacts, from slaps and kicks to caresses and cuddles.

These actions provide immense insight into their social lives and individual relationships. Two particular and specialized areas of communication by touch – sexual acts and the mother-baby bond – are dealt with in chapters 8 and 9.

Grooming

One of the most important ways that great apes send messages and convey information by touch is through social grooming. During this process, the groomer (the individual intending to do the grooming) approaches his or her intended recipient, perhaps using sounds, gestures, and physical contact (like the hand clasp mentioned below) to establish intent and initiate the session. Having settled into position, the groomer usually employs one or two hands to move the recipient's hairs against their lie of direction, exposing the lower shafts and bare skin. With a concentrated gaze, the groomer then uses the free hand, or, if both are already in use, the mouth, lips, and teeth, to delicately comb out and pick off all kinds of debris such as skin flakes, scabs, bits of dirt, and parasites such as ticks and lice. Much of this debris is then eaten by the groomer. In the social great apes – the chimpanzee, bonobo, and gorilla – this form of mutual contact is more than just a helping hand for maintaining hygiene. It is also loaded with social significance, acting as a hub for group communication. Who grooms whom, when, and for how long is enormously important.

Grooming is very common among gorillas, but it tends to be self-grooming rather than social grooming. Mountain gorillas in particular groom themselves a lot, in order to keep their long hair shiny and knot-free. That said, while social grooming is less common than among chimpanzees or bonobos, it does still occur and carries a great deal of significance, being one of the ways in which gorillas can ease tensions between one another and relax as a group. Social grooming is most common among gorillas between mothers and their offspring. Adult females may groom each other, especially if they are related. They may also attempt to groom the silverback, but he is choosy about who he allows to become so intimate with him, and he rarely returns the compliment. Sometimes females offer grooming to a blackback male, perhaps as insurance for the future when he might graduate to silverback and lead the group.

As with the other social great apes, young gorillas indulge in physical contact via play and rough-and-tumble. They push and shove with hands and feet, back into playmates, somersault onto them, and roll around on their backs. Adult gorillas, even the silverback, seem remarkably tolerant of this horseplay.

The grooming hierarchy

As befits the chimpanzee's fast-moving, fluid social life, social grooming plays several roles. It helps to sort out an individual's position in the dominance hierarchy, strengthens alliances between close friends, eases tensions by submission, and transforms aggression into calming behaviour. In general, males groom each other most often. Grooming is particularly common among allies, to reinforce coalitions. High-ranking males are usually groomed by their subordinates, rather than vice-versa. Sometimes rivals groom each other to pre-empt conflict and diffuse tension between them. Close bonds between females are less common in chimpanzees than bonobos, and so female-female grooming is too; when it does occur it tends to be between related adult females such as sisters.

During a long relaxation session, chimpanzee grooming 'clusters' can build up to include more than 20 individuals, although adult males tend to prefer smaller clusters than females do. The grooming itself is accompanied by gestures and signals such as tooth-clicking and lip-smacking. Overall, grooming is the most important social behaviour among chimpanzees.

Chimpanzees also use other forms of physical contact, such as touches, hugs, and embraces, often in ways that we instinctively understand. After two chimpanzees fight, the 'loser' takes to scratching and grooming itself, which indicates stress. However, if another chimpanzee offers contact, from a polite hand to a hug, the loser's self-grooming and scratching are much reduced, indicating reduced stress levels. We might draw the parallel of a friend offering sympathy and consolation after an argument with the boss.

Physical contact is used along with facial expressions, postures, gestures, and sounds to communicate many messages and emotions.

Close for comfort
The mother orangutan sniffs and nuzzles her baby reassuringly, each recognizing the other's scent.

Child in arms
As the mountain gorilla group settles for a rest, a female wraps her arms protectively around her infant. There can be few safer places in the rainforest.

Baby talk
As the chimpanzee mother gently strokes her worried baby, the latter becomes more relaxed and restful.

Babe asleep
A protective maternal hand over this new chimpanzee allows it to catch up on much-needed sleep.

Subordinate chimpanzees tend to approach dominant ones with submissive signals such as crouching, holding out the hand, and pant-grunts. If all is well, the higher-ranking individual will respond by touching, kissing, or embracing the subordinate.

Cultured contact

In recent years field scientists have observed an interesting form of contact behaviour between chimpanzees at Mahale, Tanzania. Known as the hand clasp, it involves one chimpanzee coming face to face with another, raising one arm, and grabbing hold of the other's hand, while using its other, free hand to groom the underarm area of its associate. The hands are joined above the two chimpanzees' heads to form an 'A'-like shape for mutual support. Usually the two individuals are male

and female. Some chimpanzees handclasp-groom every two to three hours, making this a relatively common form of behaviour. Intriguingly, one of the chimpanzee groups at Mahale clasp hands palm to palm, but in another nearby group, individuals instead press or grasp wrists. Even more intriguingly, chimpanzees who move from one group to the other forgo the old style of hand clasp and take up the style of their new, adopted group. This suggests that the behaviour is cultural, passed on through imitation and learning, rather than genetically inherited.

I'll scratch your back …

Another form of physical communication among chimpanzees is the behaviour known as social scratching. Social scratching has been recorded at Mahale and Kibale, in Tanzania, but not elsewhere. It usually occurs as part of social grooming and like social grooming seems to be linked to the social hierarchy. Higher-ranking males tend to receive more scratching from their subordinates than the other way around. The scratching usually occurs on the back of the body at the end of a grooming session. A variation on social scratching, known from Kibale, is 'directed scratching', a behaviour male allies use to strengthen social bonds. During a grooming session the recipient scratches himself in an obvious, exaggerated way on a certain part of his body. In about two-thirds of cases the partner then moves his grooming attentions to this scratched area. This has been interpreted as an example of 'referential communication'. Instead of the communicator sending out messages about his own emotional state and intentions, without necessarily expecting an outcome, he refers or directs something to the receiver's attention, with the possibility of a response. Referential communication relies on an understanding within an animal that other individuals have their own awareness and mental capabilities. We use referential communication naturally whenever we point at an object to direct another person's attention towards it. Captive chimpanzees have learned to do this, too, by copying their human keepers. Directed scratching is one of the only examples of referential communication known in wild chimpanzees, although other forms of referential communication may well exist in unobserved troops.

Tactile behaviour in bonobos

Like chimpanzees, bonobos use a wide variety of touches to communicate their moods and wants within the group. An individual who extends a hand to touch another, usually on the back, head, or rump, usually does so as an appeasement gesture of submission. A gentle pat-like touch from a dominant bonobo, on the other hand, can reassure a distraught subordinate.

Social grooming in bonobos is very widespread and linked to social rank, but it works in a different way from social grooming in chimpanzees. In the latter, male coalitions are dominant, but in bonobos, the females are effectively higher-ranking individuals. As a consequence female bonobos socially groom each other more than male bonobos do – the opposite situation to that seen in chimpanzees. There is also more grooming that takes place between the sexes in bonobos than chimpanzees, especially between mothers and their sons, and sessions tend to last longer than they do among chimpanzees.

Another type of physical contact that takes place among bonobos is collaborative genital rubbing – part of their highly sexual social life. Unrelated female members of closely bonded groups rub genital areas, a behaviour usually instigated by a lower-rank female to defuse tension and restore friendly relations after a dispute.

Friends for now
In a gesture of reassurance, a senior chimpanzee offers a finger to be nibbled by a lower-ranking individual. If the senior chimpanzee should get into a hierarchy dispute, it may be able to count on the nibbler as one of its supporters.

Mass groom-in
Grooming is 'social glue' for many apes. In this tangle of faces and limbs, young chimpanzees get to know who's who by touch and scent.

Social Life

Generally speaking, among great apes, the orangutans are the least social species. Gorillas are slightly more sociable, but they enjoy relatively limited and subdued communal lives. Chimpanzees and bonobos are more active, with more complex societies, while humans are the most sociable of all. However, although many of us now live in huge cities, each individual citizen is still tribal at heart and only has a small social network of people they know personally. These small tribes interlock to create our vast urban super-tribes. In *The Naked Ape* I pointed out that to find the natural size of the human tribe, it is only necessary to count the number of names in the average personal address book. I guessed that the figure would be under 100. Some years later, anthropologist Robin Dunbar put the figure slightly higher, at 150, which has become known as 'Dunbar's Number'.

Family matters

Compared to most other mammals, primates have exceptionally varied social lives. Five main types of social groupings are recognized. The simplest of these is the unit formed by a mother and her baby, living together but separate from others of their kind. Among the great apes this type of social grouping is confined to the orangutans.

Next in terms of size and complexity is the monogamous family group, with a male, a female and their offspring. This type of social grouping is seen among various prosimians, such as tarsiers and pottos, as well as some New World monkeys and the gibbons. The third type of group is the polyandrous family, where one female has several male partners. This social system is seen in various tamarins and marmosets. In the polygynous system the roles are reversed – one dominant male lives with several females and their young. This is the norm among gelada and hamadryas baboons, langurs and gorillas. The fifth set-up is the multimale-multifemale group, as seen among chimpanzees and bonobos. These great apes complicate matters, however, by living in groups that frequently change in size and composition – so-called 'fission-fusion' societies. In chimpanzees the main core or hub of the group is a set of males, while in bonobos, a core of females predominates.

Orangutan behaviour

The mother-young relationship of orangutans normally lasts from six to eight years, longer than that of any other wild animal. An older sibling may still be around as an infant begins to venture from its mother and their socializing can be important, although short-lived. Most adult female orangutans have established home ranges. These ranges often overlap, but the occupiers rarely come into close contact. They prefer to avoid one another and tend to move to other parts of their range to steer clear of their neighbours. Adult male orangutans are even more antisocial. They set up large and exclusive non-overlapping ranges, each encompassing the home ranges of several females, and are aggressive towards any mature male intruders (see page 205). Generally, encounters between male and female orangutans are brief. However, sometimes pairs form a consortship. An adult male and an adult female, sometimes with offspring, travel and feed together for anything from a few days to several weeks. During this time the pair may or may not mate before resuming separate lives.

Rare gatherings

The only time orangutans are seen in numbers is when they are feeding or travelling together. Feeding aggregations form in fruiting trees and are often made up of subadults – individuals past the juvenile stage, but not yet fully mature or with established ranges in their patch of forest. The various orangutans do not really socialize but tolerate one another's company as long as there is plenty of food to go around. Travel bands move along together when searching for food, perhaps for a few days at a time. In some areas the same individual orangutans regularly form travel bands and feeding aggregations, but such relationships are generally loose and short-lived.

Compared with other apes, orangutans are self-sufficient loners. The reason for their solitary lifestyle is primarily to do with food. Orangutans are mainly vegetarian and must eat a large amount of food every day to survive. Although the habitat they live in is rich, the fruits they prefer are often scattered and do not always occur in sufficient quantities to feed more than one ape at a time.

You scratch my back (Previous page)
Chimpanzees indulge in social grooming at the Gombe Stream site in Tanzania. Grooming is 'social glue', a major behavioural part of establishing and maintaining relationships within the group.

Neighbourhood watch
Grooming within bonobo groups is a perfect time to observe often-changing array of friendships – who is cosying up to whom and which relationships are on the up or fading away.

The strongest link
Female mammals – as exemplified by this devoted Sumatran orangutan mother – feed their offspring on their milk, necessitating long periods of intimate contact.

Living in numbers

Most people have heard of the term silverback. It is used to describe a fully mature adult male gorilla, who is usually 12 years of age or older. The name comes from the distinctive pale, grey, or silvery hairs on the back, flanks, and sometimes rump and upper thighs. Silverbacks also have much larger canine teeth than females or juvenile males do. These are another sign of maturity.

The silverback is lord of his group. He leads them from place to place, decides when to stay and when to go, chooses feeding sites, and defends the group from any and every threat – rival silverbacks, predators, curious human intruders, and poachers. The silverback also keeps order in his group, refereeing disputes and quashing rebellions in the ranks. His penetrating stare is often all that is needed to quell squabbles and restore the status quo. The silverback may be helped by one or more blackbacks – younger but nevertheless sexually mature males of between 8 and 11 years of age. Blackbacks provide back up to their leader's authority.

Gorillas sometimes form groups that are all male or all female. These are usually made up of subadults who have not yet joined an established gathering. By far the most typical gorilla group, however, is the harem of adult females (generally two or three) with their offspring of various ages (sometimes including blackbacks), led by a silverback. This type of group functions like an extended family, although its members are not all from the same bloodline. The adult females are usually unrelated.

Group size in gorillas

Gorilla group size varies according to circumstances such as food availability, the type of habitat, and the number of groups in a particular area. In mountain gorillas some long-term studies report average group sizes of about eight to ten individuals. Larger groups are not unknown but they are rare. One silverback was recorded leading a group of more than 30, but such groups, when they do occur, are difficult to maintain and often short-lived. Large groups tend to split into smaller ones, usually with each gaining a silverback that has been hanging around in the area waiting for his chance.

Like mountain gorilla groups, western lowland gorilla groups vary greatly in size, but on average they tend to have between four and ten members. Eastern lowland gorillas also follow the typical pattern of one male and several females with their offspring, but on average their groups number around eight to ten. In both of the lowland subspecies, multimale 'bachelor' groups may form for short periods, perhaps comprising father and sons. Very occasionally male-female groups with two silverbacks have been sighted.

Female hierarchy

Field researchers are dedicated people but estimating numbers of gorillas is very difficult. Gorillas are alert, wary, and reclusive apes, able to disappear almost silently into the dense forest where they live. Following them is dangerous – when vision is so limited there is always the risk of confrontation with the silverback. As a consequence, gorilla groups are often studied largely by the traces they leave. Common methods of gauging group size after they have gone include counting their night nests and studying their droppings, the size and quantity of which are related to the size of the individual that made them.

Females within gorilla groups generally keep themselves to themselves and may even show antagonism towards each other, unless they are blood relatives. They organize themselves into a hierarchy that depends partly on when each of them joined the group. If a lone silverback gathers a new group around him, the first female to join will usually be the most senior, and she and her offspring form a closer attachment to him than later arrivals do. Grooming and huddling are very important in maintaining the female hierarchy, and the first female grooms the silverback more often than the other females do. In every gorilla group there is a certain amount of competition between the females for the silverback's attention, and this shapes the group's social structure by encouraging some females to leave. Female gorillas use social bonding with the silverback to commit to their group.

The chimpanzee community

Following the ins and outs of chimpanzee society is fascinating but exhausting. These lively, extrovert, highly social apes live in loose, often-changing groups called communities. A chimp community occupies a home range and is centred on an alliance or coalition of adult males who associate closely with one another – rather like human 'best mates'. These chief males sort out their pecking order and continually reaffirm their relationships by various strategies, such as staying close together, grooming each other, and sharing meat after hunting. Of course, there are power struggles. A male might cultivate the favours of several others to garner support in a bid to become the most dominant, alpha male. Even if he does not succeed, then one of his allies might, which then brings him into a more senior position.

Quality time
Mother chimpanzee and baby take time out from group action to rest and restore. These quiet moments are a valuable way to reinforce their bond.

Leading from the front
A silverback western lowland gorilla (on the right) leads his extended family out into a rainforest clearing or bai in Odzala National Park, the Democratic Republic of Congo.

Mixed grooming
Female and young bonobos hold a grooming session where the youngsters learn about the etiquette of this vital activity.

Senior males have the first pick of food and females. Chimpanzee communities vary in size. Small ones may number less than 20, while larger ones can exceed 100 members. The adult males are dominant over adult females, even through the latter may outnumber them by three to one. Adults of each sex dominate adolescents of their own sex.

Fluid structure

So far, so straightforward. However within the chimpanzee community, various subgroups and parties form for different purposes such as feeding, grooming, sexual matters, social contact, travelling, hunting and defending the home area. The whole community may come together only rarely, such as at a particularly rich food source. Then they are off again, splitting up to rove in another series of subgroups, usually with around three to six members, keeping in contact with the rest of the community by loud calls. This fluid, dynamic social situation is known as the fission-fusion aspect of chimp society, and change can be rapid and without warning. One day the alpha male may have exclusive control of the breeding females, but the next his fair-weather 'best friends' might have ganged together and usurped him.

Within a chimpanzee community, the females spend most of their time with their offspring or with other females. Sometimes a group of females bands together against a new female arrival or a harassing adult male. Mothers with young also tend to associate with each other and allow their offspring to play together. Neighbouring chimpanzee communities tend to be hostile to each other. During boundary disputes, intruders are often attacked and killed, and infanticide and cannibalism are not unknown.

A matriarchal society

Like chimpanzees, bonobos are extremely social. They form mixed communities or tribes of males, females and young, which often split into smaller mixed-sex groups of normally three to six individuals, occasionally more when food is plentiful. These groups come together or fuse at large food sources and to make nests for their night sleep. Then they split apart again, perhaps to travel or forage or rest, changing their group composition seemingly at will. This is the fission-fusion aspect of their lives.

However, there is one great difference between bonobos and chimpanzees. Whereas chimpanzee society is male-dominated, in bonobo communities the females are in charge. Like chimp males, bonobo females form close bonds with their 'best friends'. They build alliances or coalitions, often including females who have arrived from other communities (which again contrasts with chimpanzees) and indulge in planning and politics to get what they want. In this matriarchal society they are dominant over the males. Within the male hierarchy, an individual's position is related partly to the seniority of his mother. If she is high up in the female rankings, he is usually senior in the male order. This can work the other way too, with a mother gaining position as her son matures and moves up the social ladder. Despite their equal complexity, social hierarchies among bonobos are not as important or prominent in the social systems of the species as they are in the social systems of chimpanzees or many other primates.

Bonobo 'free love'

A greatly publicized feature of bonobo society is its sexual nature. Sex is used freely as a social 'currency' in all kinds of interactions. Sexual activity takes place between adults of different sexes and the same sex, and between old individuals and young. Touching or rubbing genitals, the use of hand or lips, mounting from different positions – practically anything goes. That said, most sexual activity among bonobos takes the form of genital rubbing between females. Males sometimes hang from branches and indulge in 'penis-fencing'. In most animals sex is solely for the purpose of reproduction, but in bonobos it is also used for social climbing, forming alliances, demanding or repaying favours, gaining access to food and a host of other purposes.

Typical groupings

The great apes have a wide variety of social systems, from the often solitary orangutans, through the harem-like gorilla assemblage with one chief male, to the male-female pairings within an extended family of humans, to the much larger chimpanzee and bonobo communities that fragment into fast-changing smaller subgroups and parties.

Gorilla
One adult male (silverback) with 5–15 adult females and offspring, in some circumstances there may be more than one adult male

Orangutan
Mature males generally solitary; possibly small-group juveniles; mature females lone or in small groups each with one or two offspring, no long-term male-female bonds

Bonobo
Party size 3-8 within a community of 50–100-plus, no long-term male-female bonds

Human
Male and female have longer-term bond usually with other relatives in 'extended family' plus relatively few close friends

Chimpanzee
Party size 3-10 within a community of 20–100, no long-term male-female bonds

Apes alone

A great ape alone that is not an orangutan is a relatively unusual sight. Gorillas, chimpanzees, and bonobos have extremely strong social tendencies and seek out the company of their own kind. But sometimes they have problems integrating into a group, and more rarely they are actively excluded from a group they were once part of, even if they previously held an established position in it.

Although orangutans are chiefly solitary, recent insights into their lives have revealed new information about their social habits – especially the effect of food availability on socializing. The Suaq swamp area of Sumatra is a productive place for orangutans, with plentiful and regular fruits and other food supplies. In this area the orangutans are less solitary than elsewhere. Females in particular come together not only to feed but also to observe each other and allow their offspring to play together. It seems that here there is a very loose form of local community, with individuals preferring certain neighbours to others.

Transitional phase

Young male gorillas reach maturity at around nine years of age. At this time they typically leave their natal group (the one into which they were born) of their own accord, or are driven out by the silverback, to spend some time either alone or with other young males. Young males are especially likely to leave if the silverback in their natal group is himself relatively young and set for a long reign. When this is the case, becoming a blackback 'second-in-command' is a poor strategy, as any younger male would have many years to wait before getting the chance to take charge.

Lone young male gorillas roam widely, making loose contact with various groups and checking out the silverback of each one to assess the chances of a takeover. Remaining alone seems to be more common among young males than forming bachelor groups. For example, during a three-year study of western lowland gorillas at Mbeli Bai in northern Congo, 14 groups were observed with an average of eight members each, and also seven solitary males, but no bachelor groups.

During his wandering time, the lone male collects and stores information about each local group's whereabouts, for future reference. He may start to try to lure away females from an established group in order to begin his own or he may even attempt a kidnap in which a female's infant might be killed. As he gains experience and further maturity over the next three to five years, eventually his time to rule a group of his own usually arrives. He ends his solitary phase by acting as the nucleus for his own newly assembled group or as the new leader of an existing group. At the end of his reign, however, a deposed silverback usually leaves the group he previously ruled to roam alone once again.

Social exclusion

In chimpanzees, as usual, the situation is much more complex than in orangutans or gorillas. Adolescence and subadulthood are awkward times when a young male chimp, having been well tolerated by other members of his community as a juvenile, must start to find his feet. The existing alpha male and his cohorts may well view the maturing individual as a threat. The young male might then take himself out of the core area of the community to more peripheral sites, and start spending time alone. However, he does not leave the group entirely but takes every chance to watch the dominant males and study their behaviours and relationships as they feed, socially groom, challenge each other and then reconcile. Sometimes he may imitate their actions, but always at a safe distance away, from the sidelines.

Gradually, however, the young male's semi-solitary spell ends as he begins to pluck up the courage to re-integrate. This can be a dangerous time and physical encounters are common, with scratching, biting and wounding often taking place. Sometimes a male suffers repeated bullying and attacks at the hands or teeth of coalition partners – one of the reasons why adult male chimpanzees scheme to gain allies. Putting up with this sort of abuse is usually worth it, however. After a time the young male is usually taken into the fold and assumes a rank according to his abilities, both physical and political. Fighting turns to reconciliation and a new order is established. This makes sense to all group members from a survival point of view, in that the ruling males are better able to defend their area and females against rival groups when working as a cohesive team than they are when continually squabbling among themselves.

Once a male is let into the adult 'club' he is unlikely to spend further time alone, except perhaps when he nears the end of his life. Old male chimpanzees tend to become more solitary. Although they remain part of the group they spend far less time socializing and reaffirming bonds than younger males. It is almost as if, with their best years behind them, they retire and settle for a quiet life.

Social climbing
(Previous page) Female and young chimpanzees communally snack in the branches, where they are less distracted by the machinations of the males below.

Life in retirement
Silverbacks who have been displaced from their groups, like this male mountain gorilla, often spend their later years in solitude.

Who's in charge?

The aggressive charging display of a male silverback gorilla is a terrifying sight, never forgotten by people who experience it. As with many aspects of great ape life, it involves ritualized elements of behaviour – those that have been adapted and modified from their original role to serve as communication signals. The display's function is not to end with a fight but to avoid one. It is given not only to rivals who challenge the silverback's rule over his group, but in various situations that involve agitation, worry, anger, or excitement, including the approach of humans.

The full display begins with warning hoots that start slowly and gain speed. The silverback then grabs vegetation as if to feed, but instead tears it up and thrashes it around with great show and noise. Next he stands up and continues to tear at and bash the vegetation, before kicking out with one leg. Then there follows a phase of running about noisily, either on all fours or up on two feet, still attacking plants and breaking branches with great gusto, accompanied by low roars, screams and other intimidating sounds, and teeth-baring to display his considerable canines. During this there is usually a chest-beating phase where the silverback cups his hands and slaps his upper chest to make a thumping sound, amplified by the laryngeal air sacs under the skin there. The display ends with the silverback thumping the ground rather than his chest, all the time watching to see the reaction of his 'opposition'.

Usually by this stage most intruders will have sensibly retreated to safety. However, a rival silverback intent on taking charge of the group may have other ideas, especially if he is young and healthy while the resident silverback is beginning to age. In this case the challenger responds with his own display, and the two males size each other up. If they both consider they have a chance, a physical battle may ensue for leadership of the group. All the time the females watch and wait, staying well clear of the males as they charge to and fro. Should the challenger win, he may well kill any infants that are in the group, as they do not have his genes. This act will shorten the time until he becomes a father himself, since females that are no longer suckling come back into reproductive condition.

The exchange of power varies between gorilla subspecies. Among mountain gorillas, a younger male may take over a group from the silverback, his father, especially if the son was born while the silverback was middle-aged and matures as his father's power is fading. However, if a silverback dies from accident or disease (or human poachers), usually the females separate and go their own way to look for a new male who will lead and protect them. In eastern lowland gorillas, in contrast, a resident male 'inheriting' the group from the silverback is much rarer. Also, if the silverback dies, the females tend to stick together and wait for another adult male to adopt them. Staying as a group may reduce the chances of the females, and especially their offspring, falling victim to predators, chiefly leopards.

The alpha male

As described previously, the core of a chimpanzee community is a group of males who have very complicated interrelationships. The overall boss is the alpha male. He gets first refusal of food, mates, resting places, and other desirables. However, with these rights come responsibilities. The alpha male organizes his group's defence against intruders, such as rival male chimpanzees from a nearby community, and against predators and other dangers. It is a risky business leading from the front. The alpha male may also be called upon to keep the peace by intervening in arguments among other group members, including between subordinate males. He gets in quick with gestures such as slapping the ground, raising his arms, or a piercing direct stare, to threaten and subdue subordinates before they become aroused enough to think of mounting a challenge to his authority.

Usually the alpha position comes to a male of 20 to 25 years of age. The alpha male chimpanzee may not be the biggest or strongest in the group, or the best fighter. Instead, he is usually the most socially active and politically astute. He befriends some individuals – females as well as males – who then support him in his power struggle with other males to gain the coveted top spot. He shows his caring, sharing side and distributes favours such as social grooming, keeping some of his colleagues 'sweet' but keeping others on their toes, for example, with aggressive actions that suddenly turn to reconciliation. Adult male chimps are up to five times more likely to groom each other than are

Face to face
Sexual favours, including face-to-face copulation, are part and parcel of the bonobos' hierarchical manoeuvring.

Facing up
Two mountain gorilla silverbacks confront, each intending to take charge.

Face off
A mountain gorilla silverback chest-beats and lunges towards his fleeing opponent.

About face
Bonobos utilize many sexual positions as they curry favour with potential allies.

females to mutually groom. They are also many times more likely than females to greet each other by kissing, embracing, or pant-hoots and similar sound signals. However, the alpha male maintains his aloofness by rarely pant-grunting or pant-hooting to other males. He shows his status partly by his appearance, hunching his shoulders and erecting his hair to make himself look larger. He is also careful not to be too rough when demanding mating favours with females, since they become wary or even fearful of bullying males and are then less likely to offer their support.

Chimpanzees are well known for their hunting parties. The meat share-out at the end is a prime opportunity for an alpha male to dispense favours. He is generous to his loyal allies, especially the supportive females, and also to middle-rank males who are not likely to thrust their way past him to the top. He makes sure that his potential rivals are not included and that his friends are rewarded for giving him their backing.

Male chimpanzees vary in their ambitions and intent. Some have the lust for power and work hard to force their way up the dominance hierarchy to the alpha position. Others start out keen to improve their social standing but relinquish their ambition if they run into problems and resign themselves to secondary status. Still others do not seem especially concerned about their place in the pecking order at all.

Female leaders

Bonobos prefer sex to conflict. In their society both sexes have their internal pecking orders, but females generally rule over males. Bonobos are less confrontational than chimpanzees. Occasional female-female conflict occurs but it tends to be quickly glossed over, before making up takes place, generally with some form of sexual contact. Males also indulge in sexual contact following any type of tension or disagreement.

The position of the bonobo alpha female as overall community 'queen' is equivalent to the chimpanzee alpha male. Bonobo males have their own hierarchy, but females have a great say in its composition. The 'sisterhood' of leading females may band together to keep a male away from them or from food. In fact they may even attack a male they see as undesirable. Therefore a bonobo male must cultivate the females since he cannot achieve high status without their support. His own success at gaining rank is partly correlated with the status of his mother, and vice versa. In some bonobo groups the alpha male may be the alpha female's son.

Seeking reassurance
At Gombe, 12-year-old male Faben approaches female Melissa (top), who is concerned for her offspring nearby. Melissa extends a hand as a plea of friendship (middle). Faben responds with a down-turned hand, signifying assurance (bottom).

Social climbing

Most of us have friends who are pushy and ambitious, always striving and trying for some distant goal. Others we know are more relaxed and accept what life deals out to them. Raw naked-ape ambition is blamed for many problems, such as anxiety and other mental pressures, and physical illnesses associated with stress. Among chimpanzees, similarly, by no means all males develop a yearning for the alpha position. Years of study at Gombe and elsewhere have revealed that they vary greatly in their ambitions and intent.

Some male chimpanzees have the lust for power and work hard to force their way up the dominance hierarchy. Other individuals start out keen to improve their social standing, but if they run into problems, they relinquish the ambition and resign themselves to secondary status. Still others do not seem especially concerned about their place in the pecking order; they hardly ever get involved in tussles for dominance.

When a male chimpanzee rises to boss, his approach and tactics can vary widely. For example, he may accept a one-on-one challenge by another male who is larger and more powerful, only if he has the support of his back-up crew. Other alpha males avoid altogether direct challenges by bigger, stronger rivals. Instead they shift their attention to less senior males who they already know they can dominate. This they do often with a great display of bluster and show, as the other males look on. In this way the alpha male demonstrates to potential usurpers that he is in charge by bullying lesser opposition.

Divide and rule

A mixture of intimacy and intimidation seems to work well for some alpha male chimpanzees. There are close friendships with males who are not especially ambitious, to keep them on side, alternating with threats to those who might be planning a spot of social climbing. The experience that comes with age is needed to weigh up these situations. The alpha's top-spot time depends not only on his physical prowess but, as his bodily condition declines over the years, on how long he can maintain coalitions and allies.

There are relatively few instances of playful adult apes that seem to be having fun and enjoying themselves. When this does occur, on closer examination it seems some of these instances could also have social significance and a wider purpose. The questions then arise: are these chimpanzees playing in the way we understand it, and indeed what is play?

Just for fun?
Two young mountain gorillas test out their strength, speed, and intimidatory skills in a non-threatening social situation.

Patience is a virtue
The silverback gorilla, despite his power and short fuse in some situations, is usually tolerant of bothersome youngsters.

Playing

Nearly all young great apes indulge in vigorous bouts of playful activity. Their joyous, gymnastic tumbling and mock-fighting is so reminiscent of the play of our own children that it can make the most studious scientific observer smile.

Play is easy to recognize but difficult to define. The big difference between the play of the wild great apes and that of humans is that, where there is a marked decrease in most apes' playfulness when adulthood is reached, many humans play throughout their lives. Adult human play has been the secret of our success, leading us into the areas of innovation and invention that have made us the most advanced species on Earth. Lack of adult play in great apes may be one reason they have hardly progressed in areas of experimentation and innovation.

The function of play

Functionally, the play of young apes is primarily self-exploration. They are testing out the limits of their physical and mental capacities, investigating how their growing bodies work and how they interact with the environment around them. Characteristically, a bout of play involves a great deal of variation with no end goal. Play-fighting, for example, is not serious – there is no real winner or loser. Each playing ape will switch its role from winner, chasing a rival, to loser, being chased by a rival. However the games they invent will enable them to experience the sensations of being winner or loser, with all the accompanying physical movements and actions. This has sometimes been referred to as 'practising' adult activities, but it is not a specific training process. It is rather more of a way of keeping the brain active and alert during a phase of life when, thanks to parental care, the young are being shielded from the struggle for survival that they must face later on, when they themselves become adult. For convenience, students of animal behaviour identify distinct categories of animal play. For example, object play involves a single animal with an inanimate object such as a stone, twig, or pile of leaves. Locomotor play involves solo gymnastic bouts of climbing, swinging from branches, jumping, leaping, and rolling over. Social play involves physical interactions between individuals. And social locomotor object play combines all of these.

Social interactions

In social play, young great apes wrestle, trip, fumble, pull hair, bite, scratch and tickle. They make breathy giggles, chuckles, and laughing sounds that are strikingly similar to our own play sounds. Observing chimpanzees and their social interactions over the years shows that individuals vary in their degree of playfulness, from being boring non-participants to fun-loving joysters. For example, one of the more playful characters climbs into a tree. Others follow. They begin to hang, swing up and crash down. Sometimes they 'hang-grapple' or 'hang-wrestle' as each individual suspends by one hand and pushes or pulls the other.

Learning the rules

In the play behaviour termed 'circling', two chimpanzees move in a circle around each other, sometimes somersaulting, sometimes reversing direction, and sometimes breaking off to wrestle or pretend-fight. Twigs, leaves, stones, and other items may become involved. One chimpanzee throws or brandishes the object at another to instigate a session, as 'play start'. Two or three chimpanzees may also try to hold or possess the item and indulge in pretend competition for it. As described elsewhere, the 'play-hit' is when one chimpanzee raises its arm as if to slap another, while the 'play face' has the top lip covering the upper teeth; both signify that what is about to follow is intended to be fun, not serious. All of these games teach the participants about social conduct, restraint, and control.

Object play
Messing about with items from the surroundings, such as young orangutans manhandling twigs ◄ or large leaves ↗↗ is referred to as object play.

Testing the limits
Provoking a reaction from adults by jumping, prodding and poking, to see how far they can be pushed, is common in great ape play – including chimps ↗ and gorillas ►.

Helping others

Humans can show not just generosity, but altruism – selfless rather than selfish behaviours and actions, which help other humans to whom they are not related and bring them benefit, with no expectation of payback or reward. In biology the concept of altruism has slightly different connotations in that the actions of an individual benefit others while imposing a 'cost' on that individual itself. These benefits and costs are measured in terms of reproductive success – the numbers of offspring an individual has perpetuating their genes.

At first glance altruism seems to oppose the notion of evolution by natural selection, where everyone is out to maximize their own chances of survival and propagate their own 'selfish genes'. Biologists reconcile this apparent contradiction using ideas such as kin selection. In a bee colony only the queen reproduces. The female workers are sterile and dedicate their lives to supporting her by maintaining the nest, obtaining food, caring for the queen and larvae, and sacrificing themselves when attacking enemies. This may seem the ultimate in selfless behaviour. But the workers are the queen's daughters and share many of her genes. By ensuring the queen breeds, they are doing the next best thing to breeding. Kin selection theory says animals are more likely to be altruistic to relatives than towards unrelated individuals. The closer the relationship, the greater the degree of altruism.

Altruism experiments

Do non-human great apes show traits such as generosity, altruism, and a sense of fair play? In the wild it is very difficult to say. An alpha male chimpanzee may appear to share food in a generous manner with other males and with females. But he may well be dispensing these favours because he expects the support of the recipients when he is challenged for his alpha position. In gorillas, if there are related females in a group, then they may have closer relationships with each other than with non-related females, and they might even form alliances, especially when competing for the best food. The silverback might then act to prevent this happening, and make sure that he himself remains the centre of attention. It is not clear whether the females' behaviour in these situations is altruistic, reciprocally altruistic, or simply self-serving.

In captivity great apes can be studied more closely. Over the years there have been several series of experiments that have attempted to show altruistic or similar tendencies. In one set-up, chimpanzees were given the choice between obtaining food for themselves or for their whole group, with no difference in effort between these options. The results suggested there was no preference for chimps to feed the group rather than themselves, that is, nothing altruistic.

Another experiment placed a chimpanzee in the position of helping a human who 'accidentally' dropped a clothes peg and obviously wanted to retrieve it. Most of the chimps involved understood the situation and helped out by passing the peg to the person, even though there was no suggestion of them receiving food or any other reward for doing this. If the human deliberately threw the peg on the floor, the chimps were less likely to help. This seemed to show that they could discriminate between needing the peg, in which case they assisted at a small cost of effort to themselves but no gain, and wanting to get rid of the peg, when they did not help. In a third set of experiments one chimpanzee was shown some food but then a doorway to it was closed, which the chimp could not open. Another chimpanzee was able to see the first chimp and the door but not the food; however, this second chimp could open the door by pulling a chain. Naturally the first chimp tried to open the door to get at the food it knew was beyond. The second chimp, seeing its struggles, then opened the door – even though that second chimp gained no reward and did not even know why the first chimp was trying to open the door in the first place. This result suggests altruism – helpful behaviour, if you like. The second chimpanzee was certainly not opening the door for selfish reasons.

A sense of fair play

Further chimpanzee experiments suggest they have a sense of fairness, but whether or not they apply it depends on the social situation they are in. In one particularly telling experiment two chimpanzees were asked to carry out tasks and then given rewards of food, but one chimpanzee received much less desirable food than the other. If the two chimps involved did not know each other very well, then the one receiving the poorer food soon lost interest and gave up carrying out the task. However, if the two chimpanzeess were very familiar with each other, from the same close social group, then the one receiving the poorer food stuck to the task so that its friend could benefit from the better food it received. Humans can relate to this situation: we are more likely to put up with unfairness to ourselves if a friend or family member is gaining benefit, than if a stranger benefits.

These types of experiments have yielded mixed and sometimes contradictory results. Researchers continue to debate whether altruism is a purely human trait, or whether chimpanzees show it, too, in which case it may extend as far back as the human-chimp common ancestor.

Funny ha ha
(Previous page) Some chimp behaviour looks disconcertingly familiar, although their postures and expressions do not necessarily correspond with ours. Here two youngsters seem to relish larking about while a female appears to register tiresome resignation.

What's mine is yours?
As two chimpanzees share a fruit, all seems well with the world. However given their intricate and fast-changing relationships, it is difficult for field researchers to track whether this is simple cooperation or something more socially loaded.

Sharing caring
Bonobos mutually investigate a tasty snack – when times are good. Such relaxed social interaction my become less common in stressful times of food shortage or local overcrowding.

Joining and leaving

Throughout the living world, inbreeding is a bad approach to reproduction. Mating between close relatives reduces the genetic diversity that aids adaptation and survival. It causes 'inbreeding depression' – where individuals are less healthy and fertile – and throws up problems such as inherited conditions. Each of the wild great apes has different social mechanisms to avoid inbreeding and keep their species' gene pools varied and healthy.

Orangutan males tend to disperse farther from the place they were born than females. As a consequence, when they become old enough to breed they are often dozens of kilometres away from their mother and sisters. Male and female gorillas both tend to move out of their natal groups (the ones they were born into). Male bonobos and chimps tend to stay in their natal groups while the females transfer to others.

Leaving home

Male gorillas reach sexual maturity at about the age of eight to ten years. As described previously, most maturing males leave their natal group and begin a period of wandering, either alone or with a few other like-minded 'bachelor' males. Less commonly, and chiefly in mountain gorillas, a male may remain in his birth group as the second or junior silverback. He may even breed with some of the females and perhaps take over the reign when the senior silverback – who could be his father – passes on. Female gorillas reach sexual maturity at a similar age to males. Often the silverback in their group is their father. To avoid inbreeding all females depart and either transfer to another established group or link up with a lone silverback who is setting up his harem. These emigrating females tend to go farther than the males.

A female does not always set off on her own for a new home. Two females may accompany each other and set up with a solitary male to start the nucleus of a new group. As the group expands, they will retain the status of senior females by virtue of being the earliest arrivals. Secondary transfer, where a female leaves her second (adopted) group for a third, is known among western lowland and mountain gorillas.

Staying put

Chimpanzee and bonobo males are philopatric – they prefer to stay in the communities where they were born. Here the onus for moving on and mixing up the gene pool is left to the females. Most chimpanzee females emigrate during their adolescent years, generally between the ages of nine and 13. This emigration can be an extended process, taking up to two years as the female moves out of the group and then back into it again, unwilling to decide. Not all females do end up leaving their natal groups.

If a female chimpanzee finally does decide on a new group, she may have a long and arduous task fitting in with residents. Most males welcome her arrival, but she is usually younger than most of the established females, and has no offspring to aid her status. Some females may band together to shun or even reject her. In contrast, if she stays in her natal group, she usually has the support of her mother, and her integration into the adult breeding part of the community is less troublesome.Typically a female chimpanzee moves 5–20km (3–12 miles) from her natal group. Secondary transfer is rare but does happen. Sometimes a young male may accompany his mother and settle into the new group.

Bonobos have a similar pattern of emigration to chimpanzees, which means that males in a community tend to be related while females are not. Bonobo females may spend even longer associating with and 'sampling' nearby communities before they decide which one is for them. Once settled, the new female ingratiates herself with sexual acts to join the other females.

↑ Who's there?
As bonobo groups come within calling distance, individuals listen not only for potential rivalry but also for a possible opportunity for females to swap social setting.

↖↗ ...I am!
Returned calls may help to establish whether there is a social opening nearby (above left). Times for this type of auditory interaction include dusk (above right), as many other day-active rainforest animals fade into silence.

→ Making a move
The small size and family relatedness of gorilla groups mean that most females and males move on as they reach breeding age. Males display and intimidate each other as they attempt to found their own dynasties.

Territories

We like to relax in comfort at home. If we spy a stranger in the garden, the natural reaction is to get ready for a confrontation. Who is it, what do they want, and why have they come onto our private property, uninvited and unknown? The urge to defend our territory is strong.

Nature lacks wall and fences to define boundary markers, and for the non-human apes, the situation is complicated by different kinds of home. The 'home range' is an area roamed regularly by an animal or group, which they know well and use for resources such as food, water, shelter, and breeding or nesting sites. In addition there are 'core areas', usually towards the centre of a home range, where the animal or group spends most of its time, with the peripheral sections visited less often. A home range or core area is not usually defended against intruders of the same species. Active defence by calls and noises, scents, visual displays and perhaps eventual physical combat denotes a territory. The territory is very precious as regards resources. Having one might be the only way an animal can breed: no territory, no mates and so no offspring. Home ranges and even territories are not permanent. They shift with the seasons, food availability, the death of occupiers and the arrival of immigrants.

Sharing out the forest

In orangutans, the usual situation is that mature adult males each occupy large and mainly non-overlapping areas which might be called territories. One of these areas can vary from 5 to 25 square kilometres (2 to 9.5 square miles) or more, usually depending on the quality of habitat and food availability. The male's 'long calls' proclaim his ownership, warn other mature males to stay away to avoid conflict, and help them space out through the forest (see page 172). If two males encounter each other there is usually a stand-off period when they stare, roar, shake their fur, inflate their air sacs, thrash or throw branches, and generally try to appear as big and intimidating as possible. If neither backs down it may come to a fight, and although rare, this might result in serious wounds and even fatality.

The fully mature or 'flanged' male adult has leathery cheek pads or flanges, as well as long hair, large air sacs, and other trappings of his senior maleness. Unflanged males, however, are generally younger and have not yet developed their flanges, or occupied their own areas. They tend to wander with no fixed abode, tolerated by the local flanged males as long as they keep a respectful distance and do not interfere with the females. Until, that is, one of the local flanged males dies or disappears and there is a gap to fill. Within months of moving in, the new male develops his flanges and other fully mature features, and is long-calling to announce his occupation (see page 219).

In defence of the group

It is sometimes said that a male gorilla, specifically the lead male or silverback, does not defend a territory or even a home range. He defends his group, wherever they are. Much of the data on gorilla movements and ranging have come from mountain gorillas. Usually in this subspecies each group has a home range that the silverback knows well, as he exploits different parts at different times, such as following seasonal foods. Usually, too, neighbouring groups manage to avoid each other rather than come into contact. If two groups should meet, then most frequently the two rival silverbacks display aggressively with roars, chest-beating and mock charges, as described earlier, to keep possession of their females. Compared to mountain gorillas, recent work on western lowland gorillas in the Democratic

Caller ID
Dawn and dusk are common times for the male's territorial 'long call', but any time of day will do, as females listen intently to gauge to direction, distance and maturity of the sender.

Border patrol
A group of chimpanzees sets off quietly to check the outer reaches of their home area, ready to confront and even attack any intruders from the neighbourhood – especially if there are just one or two of them.

Republic of Congo and CAR shows that groups encounter each other much more often, and also they are much less antagonistic when they meet. These gorillas have overlapping home ranges that vary with the seasons, food availability, and group size, from 5 square kilometres to more than 30 (2 to 11.5 square miles). However, relatively few inter-group encounters involve threat displays. In one series of observations at Lopé, Gabon, only 3 out of 43 encounters led to fighting. In the others, relationships varied from cool but non-threatening, to almost affable. In fact some groups even intermingled.

Genetic analysis of hair, faeces, and other bodily traces suggest why. The silverbacks leading neighbouring groups are usually close relatives, such as cousins, brothers, or fathers and sons. They probably recognize each other from their upbringing. As they left their natal groups, they did not go too far. They form a male kin network sharing family genes, and since proper fights involve serious risks, they prefer a strategy of non-aggression where possible.

Extra-friendly relations

Each bonobo community occupies a home range which the members roam for foraging and shelter. Range size varies with the time of year, amount of food available, and other factors, but usually neighbouring ranges overlap, sometimes extensively. However, predicting what will happen when two communities meet is difficult. There may be some hostile reactions, with displays and screams and barks, at least initially. In other cases these preliminary tensions are minor and quickly ease as the two sets of community members relax and socialize. Indeed some intergroup meetings end up laced with the usual bonobo preoccupation: sex. While males from the two groups tend to be more standoffish with each other, the females might socialize intimately, rub genitals, and mutually groom, while males and females may even copulate as the youngsters play together.

Communities at war

As in the bonobo, a chimpanzee main group or community frequents a home range, particularly a core area which forms about one-third of it. The range area varies from less than 10 square kilometres (4 square miles) to more than 50 (20 square miles), depending on factors such as overall group size, the type of habitat from thick forest to semi-open savannah, and food resources. However one of the main factors in home range area is the group's number of adult males, and their skills at confrontation by display and actual fighting. This has a much greater link to home range size than, say, total group size or fruit availability. Aggression between neighbouring groups of chimpanzees has been well documented over the years. Several leading males may set off to wander their area, ready to threaten or even attack any stranger chimpanzees they find. When on 'border patrol' these males move quietly and stop often to listen for unusual sounds. How often these patrols occur depends on the number of males, whether they might go hunting for meaty prey like monkeys as well as intruders, the 'neighbour pressure' from chimps in nearby communities, and also the number of females in their community who are in reproductive condition.

Chimps at war
What happens in an encounter depends partly on numbers. For two approximately matched groups the result can be aggressive calls and displays. If a larger group comes upon just one or two strangers in their area, it can lead to attack and death.

Sex Life

In terms of reproductive anatomy and physiology, we can identify many similarities between ourselves and the other great ape species, as well as a few interesting differences. But for other aspects of sex, it can be very difficult to find parallels. In many areas of wild ape behaviour we often see ourselves reflected. Our hairy cousins display a range of emotions that we find familiar – apprehension, affection, enjoyment, and rage – and they indulge in similar behaviours, such as argument and play. However when it comes to matters of partnership, long-term pair-bonding, love and romance, and understanding that sex leads to babies, it seems that we as a species are on our own.

How many mates?

The primary function of sexual behaviour, involving a male and a female, is to reproduce the species. The basic mechanisms of reproduction are the same for all ape species, but the mating patterns of the different great apes each have their own distinctive features. In humans the most common situation, taking a global perspective, is for a female and male to form a long-term pair bond and to produce a series of overlapping offspring, with births separated by one or more years.

Of course there are various cultural variations on this theme, as well as varying degrees of legalization, for different types of sexual behaviour. There are arguments that monogamy – a system with just two mating partners together at any one time – is a social, cultural, or legal institution rather than a biological instinct. Even so, it is the most common situation for humans. It is also the general situation for the lesser apes, or gibbons – but not for the other great apes.

One-man band

Orangutans tend to be polygynous – one male breeds with several different females. Male orangutans mate with as many females as they can attract from their area of dominance. Each female, as a result, may be restricted to mating only with the dominant male. However, with such a long interval between births, around eight years, a female orangutan may get to mate with more than one male in her lifetime. By the time she is ready to conceive again, the original dominant male may well be long gone and another mature 'flanged' male (see page 204) may have taken his place. Recent studies have also shown that 'unflanged' males who lurk around and between the areas of flanged males, and who have not yet risen to dominant territory-holding status, are far from idle. In fact according to one estimate unflanged individuals are about as successful at fathering offspring as the dominant flanged males. The two types of male have contrasting strategies, characterized as 'call-and-wait' versus 'sneak-and-rape' (see page 219).

Dominance without exclusivity

Most gorilla groups have one mature male leader, the silverback, who usually has a fairly secure tenure of polygyny, with several female mates. But even then, his is not an exclusive right. Most females cultivate their social and sexual relationships with the silverback, hanging around him in a form of interfemale rivalry. But rarely females have been known to slip away temporarily to copulate with another male hanging around the periphery, who is perhaps hoping to lure one or more females away with him. When this female returns to the group the silverback, in his role of benevolent dictator, seems to take little notice. In one sense then, the gorilla and orangutan situations are similar, one male mating with several females. They differ, however, in that the female gorillas are together in a group while the orangutan females are scattered in their home ranges.

In gorilla groups with more than one adult male, all of them generally indulge in sex. The dominant silverback has by far the lion's share with up to 90 per cent of matings, especially with the more mature females, those who have just arrived in the group, and those who are pregnant. This leaves younger and subadult females for the less dominant males. In mountain gorillas, if a subordinate male attempts sex he runs an approximate one-in-three risk of being interrupted by another male, usually his superior. But the harassment is generally low-key with only mild chastisement. Female gorillas, for their part, often initiate sex and may mate with all the available males in their group, although they do so more often with the silverback.

Multiple partners

Among chimpanzees and especially bonobos the situation is less restrained, to say the least. Both males and females mate with various individuals – theirs are systems of unrepressed polygamy. A female chimpanzee at her most receptive phase, when conception is likely, tends to prefer the higher-ranking males in her community to those of lower status. Recent observations of chimpanzees at Kibale, Uganda, have also shown that males tend to prefer older females, generally those that are over 30 years of age. It may seem surprising to us, but they will chase after a 50-year-old in preference to a 20-year-old female. This may be because a female chimpanzee who has reached 30 or more years of age effectively has a proven survival record. On top of that, she has probably also already raised one or two young, and so she is likely to be a good and experienced mother.Bonobos are well known for their free-and-easy sex lives, where all kinds of sexual contact occur. This is partly linked to the fact that the female bonobo's sexual condition and receptiveness is almost independent of her reproductive and menstrual cycle (see page 210). Females mate with males even when it is very unlikely that they will conceive.

No rush
(Previous page) Gorillas are generally the least sexy of the great apes. When times are good they prefer to idle and play it cool, perhaps with a spot of what we might refer to as gentle canoodling.

Big daddy
The hulking presence of the mountain gorilla silverback both deters other males from muscling in and encourages females (here two, with juveniles) to stay with him for protection and mating rights.

Jealous guy
A mating pair of bonobos is disturbed by another male, possibly a rival, 'jealous' or desiring of a senior female's favours.

Starting out

The only body system that does not work at birth is the reproductive system. It begins to mature and function at the time we call puberty. This happens under the control of bodily 'messenger chemicals' known as sex hormones, the main ones being oestrogen and progesterone in the female and testosterone in the male. In humans puberty is linked to a growth spurt in physical size and the development of secondary sexual characteristics – breast enlargement, redistribution of body fat, and a slightly deeper voice in females, facial and chest hair growth, increased muscle mass, and a breaking voice in males, along with the appearance of underarm and pubic hair in both sexes.

The pubertal growth spurt and many other secondary characteristics are less marked or absent in the other great apes. The main exceptions are the male orangutan and gorilla, both of which experience a significant increase in physical growth, so that within a few years they are up to twice the bulk of females of similar age. Technically speaking, puberty is the term used for the time it takes for the anatomical and physiological changes to take place. Adolescence, on the other hand, involves the overall transition from immature child to mature adult, involving not just biology, but also social and psychological changes – these are described on page 240.

Sexual anatomy

The sexual anatomy of the other great apes should be familiar enough – it is much the same as ours. However, all of the non-human great apes have a bone that we lack, known as the baculum, penile bone, or os penis in the male, and the baubellum, clitoral bone, or os clitoris in the female. Interestingly, these bones are possessed not only by the other great apes but by virtually all other primates and by many other mammals too. In the male the bone helps to allow for quick penetration, without having to wait for a blood-engorged erection. However, compared to those of other mammals, in the great apes these bones are much reduced for the size of the body, being only 1-2cm (0.4-0.8in) long.

The exact reason why our species has lost the os penis and os clitoris bones is unclear. It is perhaps linked to sexual selection. The theory is that a firm, erect penis is a sign of a fit and healthy male. This is because various illnesses and health problems can interfere with erection, from stress and overindulgence to diabetes and nerve system degeneration. The human male penis is also famously much larger relative to body size than those of the other great apes – the gorilla's, for example, is just 5cm (2in) long when erect. Gorilla testes are also small in proportion, but chimpanzees have the largest testes of any primate.

Female differences

Another difference between the great apes concerns the appearance of the female external genitalia, the vulva and labia, and the skin around them, the perineal area. In chimpanzees and bonobos these parts swell considerably or become tumescent, and also change colour to pink or red, as part of the reproductive cycle (see overleaf). In these two apes the first signs of puberty among females are these perineal swellings – in the bonobo they start to occur from about the age of seven to nine years, even though female bonobos do not reach full adult size until they are between 13 and 15 years old.

The onset of puberty

In the non-human great apes, the timing of puberty is varied and complex. It often sets in earlier in captive apes than it does in free-living ones. Factors such as food supply and nutrition, seasonal influences and position in the dominance hierarchy all play a part. In general, female gorillas become sexually mature at about the age of seven to eight, although some mountain gorillas have been known to become mature at the age of six. Male gorillas begin to reach maturity from around the age of eight. Both male and female chimpanzees enter puberty when they are around six to eight years old.

In the African ape females there is a period known as 'adolescent sterility' between the first showings of sexual development and the ability to conceive. This is often linked to social matters such as moving from one group to another, as we shall see later on in this chapter.

Orangutan females pass through puberty at around eight to ten years old, although, as with all great apes, timing can vary. Some female orangutans begin puberty when they are as young as six, while others may not start until they are eleven. Regardless of when they enter puberty, most give birth to their first offspring in the wild when they are about 15 years of age. Male orangutans may go through puberty at any age from seven or eight onwards, but often they do not become fully 'flanged' until they are 20 or older. The age at which this happens depends upon the opportunity or otherwise to grab a patch of forest as territory.

Ready to mate
As a pair of bonobos prepares for copulation, they are busy looking around to see what's happening with other members of their group. For these great apes sex is a very public, frequent and casual affair.

Calling out
(Overleaf). Female chimpanzees may sound out in response to a male's approaches, as part of courtship, but this is different in nature to the copulatory call that occurs during the sexual act.

Courtship

The non-human great apes tend not to go in for flowers, rings or other gifts, moonlit walks hand in hand, or adjoining back seats at the movies. Despite the human connotations of 'courtship', however, this term is accepted among biologists and animal behaviourists for the process of recognizing, attracting, and choosing a mate. It occurs throughout the animal kingdom, and often involves visual displays, sounds, and smells, and behaviour patterns, which have become ritualized – detached from their original function and modified for communication.

The main aim of courtship is to check out potential mates and make sure they are of the same species as you and to ensure that they are fit, healthy, and vigorous. In short, its role is to find a suitable source of genes for potential joint offspring.

Hardly subtle

Each type of great ape has its own courting habits. Many of these are rather 'up front' compared to the subtleties of the human situation. For example, the male chimpanzee is largely uninterested in sex unless stimulated by a 'pink' or sexually attractive female. Her colourful swellings, behavioural changes in posture and attitude, and newly produced sex pheromones are what turn him on. Now excited, he may well begin 'courtship' by opening his legs wide to display his sex organs. He might tap the ground with his foot or knuckles, swish or pluck leaves from a nearby branch, walk about in a strutting or swaggering fashion while making sure his erect penis is noticed, or reach out as though beckoning. If the female is impressed enough she will turn around in front of him to show her pink and swollen rear end, and then crouch down. Copulation will then begin. The overall impression of courting chimpanzees is that they are relatively no nonsense and tend to get on with the job..

Consortship

A slightly different strategy for the less dominant male chimpanzee is to set up a consortship. He approaches a female and displays as above, but if she shows interest, he then moves away and looks to see if she follows. If she does, the two creep away from the main socializing area and spend time on their own, intermittently grooming and mating over a period of anything from a few hours to a week or two, before rejoining the main community. This 'sequestering' of a female by a male does have risks, including greater exposure to predators and to roving bands of aggressive chimpanzees from neighbouring communities. A subordinate male is likely to be successful at forming a consortship if he targets a female early, just as she is starting to show sexual swellings. Otherwise she is occupied by the higher-ranking males.

Females take the lead

Gorillas and orangutans have a less frantic approach to courtship than chimpanzees. The female gorilla usually has just one or perhaps a couple of males as possible partners. Since it is the males who have more choice, they tend to be fairly relaxed about the whole business. It's the females who are pushier and more assertive as they come into the mid-cycle fertile phase. The female gorilla approaches the male and displays her rear, perhaps as a response to his initial chest-beating or similar signals. He enters into the spirit by beating some more and then strutting about in a stiff-legged fashion, perhaps slapping or tapping the female and sniffing her genital region. She puts on a 'tight-lipped' expression, lips together as if nervous, and crouches to reverse towards him for copulation.

The conspicuous pink sexual swellings of chimpanzees and bonobos, which signify readiness for sex, would be of little use to a female orangutan. She spends most of her time out of sight of males in the thick forest canopy. The local dominant male's awesome 'long call' regularly announces that he is present and ready to mate. As a female comes into her fertile phase she actively seeks out the male and approaches him, displaying her rear as a sign of readiness. He shows off his penis and partakes in branch shaking before the two mate. This is the 'consensual' version, which contrasts sharply with 'non-consensual' where the female resists (see page 219). As with chimpanzees, a male and female orangutan may enter into a consortship arrangement and continue to mate over several days or weeks.

Anywhere will do
Bonobos lead such sexually charged lifestyles that a small pool of water is no deterrent. Face-to-face copulation – once thought to be an exclusively human position – is well documented.

Breeding cycles

Female great apes have their reproductive parts within the lower abdomen, as in other mammals. There are important differences in the way they work compared to other mammalian species but the basic layout is similar. There are two ovaries, each of which periodically releases a single microscopic ripe egg cell. The egg moves along the oviduct (egg tube) to the uterus, or womb. At this time, if copulation has occurred recently, it may be fertilized by male sperm, and pregnancy starts. The neck of the womb, or cervix, leads to the vagina or birth canal, and this opens between the legs at the vulva with its fleshy 'lips' or labia.

The differences between the female reproductive parts of great apes and other mammals are in timing and lining. Most female mammals have a time called oestrus, when the ovaries release their egg cells. This is under the control of several sex hormones or chemical messengers. Follicle-stimulating hormone, FSH, causes the eggs in the ovary to ripen. To prepare for nourishing the fertilized egg during pregnancy, the lining of the womb, known as the endometrium, thickens with blood-rich tissues under the influence of the hormones oestrogen and progesterone. A surge in luteinizing hormone, LH, triggers egg release or ovulation. If no conception occurs the lining is then reabsorbed.

Oestrus

Many female mammals go through this process, the oestrous cycle, at certain times of year – for example, each spring, or perhaps spring and autumn. At the point of the cycle when the female is about to ovulate, or release eggs from the ovaries, she is usually most receptive to mating – in oestrus, 'in season,' or 'in heat' – since this is the most fertile phase, the prime time for conception.

Female great apes take this a stage further. Their cycles occur more often, roughly every four to five weeks depending on the species and factors such as health and food supply. Also, the womb lining is not reabsorbed but lost through the vagina as the menstrual flow (menses) or period. This version of the reproductive cycle is termed the menstrual cycle. It means that great ape females are fertile and can conceive every few weeks. Detailed studies of the female hormones including FSH, oestrogen, progesterone and LH demonstrate how similar their levels and cycles are among all of the great apes, including humans.

Behavioural changes go along with the cycle, making the female more ready to mate at the time of ovulation (in the non-human apes and, as recent research shows, possibly in humans too). In chimpanzees and bonobos this most fertile time is signalled by colour changes and swelling of the external genitals and perineal region, sometimes known as being 'in pink' or 'in oestrus'. In chimpanzees this usually lasts for 10–12 days in the middle of each approximately 36-day cycle. In bonobos the cycle is usually a day or two longer, but the pinkness is greatly extended, to 20 days or more. Bonobos are often described as being 'released from the bonds of oestrus' and noted for their use of sex for purely social reasons. In female gorillas the oestrus phase is much shorter – two to four days – and with much less if any 'pink'. Orangutans, who lack the 'pink' show, are in oestrous for about four to six days. Most other primates also have menstrual cycles, although the menstrual flow may be much less marked. The length of these cycles varies from 45 days or more in some lemurs to less than 25 days in certain monkeys.

The default condition

In humans, depending on an individual woman's situation, menstrual cycles may continue uninterrupted for years. For the other great apes, they usually do not. This is because pregnancy or lactation (breast-feeding) is the 'default' situation. Both pregnancy and lactation produce hormones that suppress the menstrual cycle. When the ape mother has weaned her offspring, the hormonal suppression is removed and she begins her cycle again. But soon she mates and gets pregnant, and that's that for several more years.

The male condition

Male great apes differ from females in their reproductive equipment and lack of sex-hormone driven cycles. The male parts are mostly outside the abdomen, rather than within. Two sperm-producing glands, the testes (testicles), hang in a skin bag, the scrotum. The vas deferens tube conveys sperm from each testis to another tube, the urethra (which at other times carries urine from the bladder). The urethra runs along inside the male organ, the penis. During sex the penis becomes engorged with blood and erect, so that it can penetrate the female vagina and deliver sperm there by ejaculation.

The chief male hormone is testosterone but there is no male equivalent of the menstrual cycle. Sperm production goes on continually, although there may be dips or peaks associated with the seasons, food supply, availability of females, and matters of dominance in the social hierarchy.

Mating

Being a female chimpanzee in sexually receptive condition can be a tiring business. She may be pursued and harassed by not one or two males but several, all wanting to copulate, which they often do, and in quick succession. The number of females in sexual condition has an effect on community dynamics. The more there are, the larger the parties tend to be for food foraging and travelling, as several males coalesce around each female.

A recent study of chimpanzees at Bossou, in the Republic of Guinea, showed that food can be used to gain sexual favours. Male chimpanzees who offered desirable food to sexually receptive females had a greater chance of mating, and especially of forming an exclusive male-female consortship. In fact male chimpanzees take the risk of raiding crop fields and plantations just for this reason. They steal fruits such as papaya, maize, oranges, and pineapples, since these impress the females most in the 'food-for-sex' trade. These foods have been likened to the human version of the box of chocolates.

Standard position
Male chimpanzees tend to get the mating act over fairly rapidly, often eager to resume their social manoeuvring. The female may or may not make copulatory calls at this time, depending partly on who is nearby.

The chimpanzee copulatory call

At the female chimpanzee's most fertile stage in her cycle, around ovulation, she shows preference for the alpha or other high-ranking individuals over subordinates. The actual copulatory act is usually brief, with a standard position of dorsal-ventral, the female crouching and the male standing behind. During this time the female chimpanzee, as with many other female primates, may well emit copulation calls. In her case it is a type of high-pitched 'ow-ow-ow' moan. Reasons for chimpanzee copulation calls have long been discussed. One proposal is that a male chimpanzee may attack and even kill an infant he suspects was not fathered by him. So by advertising herself for frequent sex, the female gives several males the opportunity of fatherhood and thereby protects her future offspring.

Recent research has revealed that in some cases the female chimpanzee makes no copulatory calls at all; this tends to happen when there are other, more senior females nearby. Presumably she does not wish to attract their attention since they might try to break up the liaison. At other times the female may be especially vocal during sex, this is often when other high-status males are nearby. Here she is attracting them for sex and spreading wider the net for possible paternity. Copulatory calls are given throughout the female's receptive period, not just when she is most fertile. So chimpanzee copulatory calls are not simply linked to the mating act, but also to the social situation.

Mating positions

As described elsewhere, bonobo sex happens 'at the drop of a hat'. It is often more to do with social communication than procreation. There are also differences in the bonobo female's sexual parts, compared to those of the other great apes, and these seem designed to heighten sensitivity. The clitoris, in particular, is more prominent, forward-pointing, and subjected to greater swelling or tumescence during the reproductive cycle. This means it is strongly stimulated by both front-to-front or ventral-ventral copulation (as well as by rubbing against the genitals of other females, see page 186).

For many years, ventral-ventral or 'face-to-face' copulation was thought to be an exclusively human trait – like tool-making or cultural transmission of behaviours. As we became more knowledgeable about bonobos, however, it emerged that they too habitually employ ventral-ventral as a sexual position, among many others. Ventral-ventral is

known in the other great apes, but in bonobos it may be used in one-third or more of all matings. In fact one of the female's invitations to sex is to lie on her back near a suitable male.

More recently the first photographic evidence has been captured of wild western lowland gorillas, indulging in front-to-front copulation. This occurred in a forest clearing or baï at Nouabalé-Ndoki National Park, in the Democratic Republic of Congo. The female involved was well known to her human observers. She had already come up with several forms of unusual and innovative behaviour, including the use of a stick to check a pool's water depth.

How long sex lasts

The sex act in gorillas is generally low-key and lasts from seconds to several minutes. According to one survey, the average time taken is about one and a half minutes. In contrast, and despite the greater frequency and fuss, chimpanzees average 10 seconds or less, and bonobos 10–20 seconds, although this is sometimes extended to over a minute. This correlates with the high-speed, socially intense lives of these apes, who have so much else to do besides actual copulation, such as forming alliances and sorting out dominance hierarchies.

In all the great apes, the mating pair may change their position mid-stream, something that very few other primates or mammals do. Also the female may look over her shoulder in the dorsal-ventral position and make eye contact with the male, and perhaps contribute her own pelvic thrusting movements; then again, she may not.

Orangutans tend to take longer over the sex act than other wild great apes, usually at least several minutes and often up to a quarter of an hour. Orangutans are also quite adventurous in their sexual positions, which include ventral-ventral, dorsal-ventral (male behind female), side-side, male on top and female on top. What is more, they manage all this while balancing on a branch or even hanging from it, and sometimes even as the female's older offspring harasses the male, presumably trying to protect his mother.

So-called consensual copulation results from the female orangutan initiating sex after hearing the long call of a mature, fully-flanged male. A complication of the orangutan sex life is that there are, in effect, two kinds or 'morphs' of males. One is the flanged male, with the fleshy cheek pads, throat, or laryngeal air sacs, longer hair growth, and other signs of full maturity. He establishes a home range that excludes other flanged males but includes several local females. The other morph is the unflanged male. He lacks the flanges or cheek pads, the other physical accoutrements, and the territory to go with them. This phenomenon of two forms of male is extremely unusual among mammals and is known as bimaturism.

On watch
Despite what seems to our eyes to be an intensely involving event, mating apes such as these mountain gorillas maintain alertness for danger and intruders in the surroundings.

Enforced copulation

The unflanged male has a furtive, low profile life. He lurks away from other males as he 'trespasses' in their home ranges or wanders the 'no-orangutan's-land' around the periphery. The unflanged male is usually a late adolescent or subadult who is developmentally arrested. His sex organs work and he can, and indeed does, father children, but his completely mature body form is not yet achieved. His strategy is to seek out a receptive female and force her into sex, sometimes quite violently as she protests and tries to escape. In a series of observations at Tanjung Puting National Park, southern Borneo, almost nine-tenths of approaches to females by unflanged males were resisted. Unflanged males are far less likely to attempt to mate with females if there is a flanged male nearby.

The unflanged male's tactics are known by various terms, including non-consensual copulation, enforced sex, and 'rape'. The strategy of flanged males is normally known as 'call-and-wait' while that of unflanged ones is 'sneak-and-rape'. When a mature male dies, a gap in the local flanged male population occurs and the unflanged local males make a bid for it. Within a few months the successful male has developed his cheek pads and other features and is long-calling for females, having changed his copulatory habits.

Consensual sex versus non-consensual sex

In a recent study of orangutans in Sumatra, researchers concluded that flanged and unflanged males had approximately equal success rates in producing offspring. However, the study period included changes among the local flanged males that may have affected the situation. One proposal for the workings of bimaturism is that the flanged male's long call is heard by unflanged males and causes them a state of high stress, where hormones such as cortisol suppress their usual continued development. This has been dubbed 'acoustic castration'. However, studies of captive orangutans of various ages suggest that developing adolescents have higher stress hormone levels than developmentally arrested adolescents or flanged males do. The implication may be that adolescents enter the unflanged stage to avoid stress, rather than because of it. Further work is needed to clarify the situation in the wild. How did the unusual condition of male orangutan bimaturism and two types of sex arise? One theory involves the prehistoric biogeography of South-east Asia. From about five million years ago, climate change – possibly the start of the cycle we call the El Niño Southern Oscillation – led to periods when trees fruited less. Orangutans and many other animals were forced to endure times of food shortage. It is possible that the original orangutan social system was similar to that of gorillas, with one male and several females living together. With the new decrease in food abundance, however, they had to spread out and forage alone. They kept the same type of mating system, but now the females were unguarded for long periods. This left the way open for the rise of the unflanged male as a 'sexual predator.'

Pregnancy

As a group, mammals are distinguished from other creatures in that their offspring undergo early development in the womb, or uterus, of the female, rather than outside the body in eggshells or other containers. Only a very few mammals, known as monotremes, lay eggs. Monotremes are confined to Australia and New Guinea and include the duckbill platypus and about four species of echidna or 'spiny anteater'.

In the great apes, as in other mammals, sperm cells are introduced by copulation into the female reproductive system and usually meet a ripe egg cell in the oviduct (fallopian) tube, after that egg has been released from the ovary. Here one sperm cell joins with the egg and fertilizes it. The fertilized egg then continues its journey along the tube and into the womb, dividing as it goes into two cells, then four, then eight and so on. A week or so after fertilization the original egg cell has become a ball of fast-dividing cells – an early embryo. This burrows into the lining of the womb, which is blood-rich as part of the regular menstrual cycle changes. Safely implanted, the embryo continues to grow, drawing nutrients to fuel the growth from the womb lining.

From embryo to foetus

The early great ape embryo grows and changes shape rapidly. Broadly speaking, it develops 'head first'. The brain starts to form very early on and dominates development for a time. Next, the spinal cord, vertebrae of the backbone and torso take shape, along with the internal organs such as heart and digestive system. The arms and legs follow, beginning as flap-like limb buds that grow and lengthen. Details such as fingers and toes start to appear later.

It is very difficult to tell apart the growing embryos of the various great ape species – they all look extremely similar. By about two months after fertilization all of the main body parts have formed, even though at this stage the tiny being is hardly larger than a grape. From this point on the developing baby is known as a foetus. With its main body parts in place, the last months of pregnancy are devoted to growing larger and stronger, and adding developmental details such as fingernails, toenails, and eyelids. The foetus is provided with nutrition and other essentials by its mother, through the placenta. This disc-shaped organ is embedded in the womb wall and will eventually follow the baby into the outside world as the afterbirth. Blood from the ape foetus flows to the placenta along the umbilical cord. Within the placenta it passes very close to pools of maternal blood (although the two types of blood do not mix). Here vital substances such as nutrients and oxygen are absorbed by the blood of the foetus and waste substances are removed. The cleaned and enriched blood then returns along the umbilical cord to the foetus.

Time and space to grow

Among land mammals there is a very broad relationship between the length of pregnancy, or gestation, and body size. The largest land mammal, the African bush elephant, carries her baby for some 22 months, while the pygmy shrew – one of the smallest – does so for only three weeks. Overlying this general trend is the tendency of primates to have longer pregnancies than their body size would suggest. Primates also give birth to fewer offspring each time than most other mammals do. In many primate species, including our own, a single baby is the norm. This contrasts strongly with the multiple offspring seen in litters of cats and dogs, for example. Being alone gives the primate baby time and space in the womb to develop to a more mature state before leaving it. Primate babies are born with their eyes open, for example, whereas puppies and kittens are born with theirs closed.

Signs of pregnancy

A typical human female gains around 8–13kg (17.5–28.5lb) during pregnancy, of which 3–4kg (6.6–9lb) is the baby and 0.7kg (1.5lb) is the placenta. Fluid in the womb and increased muscle mass account for one kilogram (2.2lb) each. Another 1.5kg (3.3lb) is added in blood – pregnant women have far more blood pumping around their bodies than women who are not pregnant. Another 2kg (4.4lb) or so is made up by other body fluids – this may cause thickening of the ankles and other signs of what is commonly called 'water retention'.

In the other great apes the weight gain shown by pregnant mothers is less, both in real terms and also in proportion to bodily size, being about 5–6kg (11–13.2lb) in chimpanzees and 6-8kg (13.2–17.6lb) in gorillas. The babies themselves are also correspondingly smaller at birth. Pregnancy can also be harder to spot in wild great apes for other reasons. For instance, the human female body shape incorporates a markedly narrower waist compared to the other great apes, who are often replete with food, may well look pot-bellied, and also stand upright much less than humans. And of course the human great ape lacks the hairy body covering of the other species. These factors mean the prime sign of human pregnancy – the enlarging 'bump' on the abdomen – is less marked in the other great apes than it is in humans. There are other signs that can give a female's condition away, however. During pregnancy the non-human ape mothers tend to show several behavioural and dietary changes: socializing less, avoiding males more, and resting longer. Gorillas have been observed to show signs corresponding to NVP (nausea and vomiting of pregnancy) or morning sickness. It is not universal and some show no signs at all.

➲ **Eating for two**
Like any expectant mother, the pregnant orangutan increases her dietary intake as gestation progresses. She must exercise greater caution in climbing. Branches that would normally bear her weight may crack under her bulk.

Birth and babies

The birth of a baby is one of the most important and emotionally charged events in life – for all the great ape species. The intense bond between mother and offspring created at this time will last for years, beyond infancy and physical dependence of the child on the mother, into adulthood and the next generation.

With great apes it is almost always 'a baby'. The vast majority of great ape births produce a single youngster. Twins are known but rare. They are probably more frequent in humans (varying from 1 birth in 30 to 1 in 300, depending on factors such as ethnic origin) than in the other great ape species. Triplets are exceptionally rare, examples being chimpanzee triplets born at Lincoln Park Zoo in Chicago, Illinois, in 1979, and another set at Alamogordo, New Mexico, in 2000. For wild great apes, only one baby from a multiple birth is likely to survive, and even this is sometimes doubtful. The supply of nourishment within the womb, and then afterwards, as breast milk, is usually insufficient. That said, mountain gorilla twins born in 2004 in Virunga survived, and another set who arrived in 2008 at Bwindi made it through their critical first few weeks.

Breeding strategies

Why do great apes usually have just one baby? All primates, and great apes in particular, are what biologists term K-strategy species, as opposed to r strategists. The r/K selection theory is widespread in natural sciences and has links to habitats, ecology, and the types of environmental niches that species occupy. Simply put, r-selected species produce many, many offspring, as eggs or young, but then devote little further time or parental care to them. They let them take their chances in the struggle for survival. Of the many offspring, hopefully a few survive and live long enough to breed. K-selected species put their energies and resources into just a few offspring and lavish on them as much care as they can, for as long as possible. This gives these youngsters a much greater chance of survival to breeding age. Among mammals across the r/K spectrum, species such as rabbits and mice are towards the r end. Elephants and whales are very much at the K end, right next to the great apes.

Impending birth

Away from evolutionary theory, it's a quiet night in the rainforest. The heavily pregnant wild great ape is ready to give birth. Depending on her species, in the past few days she has either been more reticent in socializing within her group (African species) or meeting with neighbours (orangutans). As labour approaches, she may move from her night nest to nearby branches, showing signs of discomfort and changing position often. She feels her birth opening repeatedly and inserts her hand to check what is happening, then observes and sniffs the hand for signs of fluids. Her body movements, facial expressions, and overall behaviour suggest continued discomfort (but rarely overwhelming pain).

Contractions proceed quickly as the womb muscles tighten to push the baby through the cervix (neck of the womb) and along the birth canal (vagina). The waters break as amniotic fluid is released with the rupture of the membranes surrounding the baby in the womb – the mother may facilitate this by picking at them. The actual delivery is also rapid, taking just a minute or, at most, a few. The baby's widest part, its head, emerges first, steadied and supported by the mother's hand. Its whole body follows, slipping out easily. The mother cradles the newborn baby and begins to lick it clean and pick off or suck away the birth membranes, especially around its face and head. The highly nutritious placenta (afterbirth) soon arrives. This is too good to waste, and she eats it almost at once, biting through the umbilical cord and all the time making sure the baby is safely supported against her front, guarded and warm. Within minutes she has licked the infant clean, wrapped her arms protectively around it, and is sniffing, licking,

hugging, and caressing it. She examines the new arrival intently as she begins the first of probably thousands of grooming sessions.

The brand new baby is largely helpless and cannot even grip well at first. It is the mother's duty to hold and support it against her body. However the baby's rooting reflex causes it to turn its head and seek the nipple for a first feed of her milk. The whole process, from birth to this point, has taken between one and two hours.

Back together
(Previous page) The tiny size of this new lowland gorilla is apparent as its mother strolls confidently along a well-worn track between feeding sites.

Experienced mother
Fifi, one of the chief chimpanzees at the Gombe site for several decades, holds her three-day-old offspring. Fifi had produced nine offspring by the time she disappeared in 2004, aged 46.

Birth problems

Of course, not all ape babies arrive in the outside world in this way. Births can happen during the day, and on the ground rather than in a tree. They can take less than half an hour, or more than two hours and be more difficult. Sometimes the baby is not in the correct position for headfirst delivery, in which case labour can be more prolonged and troublesome, and occasionally end in tragedy. The umbilical cord and even the placenta may be left dangling rather than being bitten off. Orangutan mothers have been observed blowing or sucking at the faces of their newborns, perhaps trying to encourage them to breathe.

Some gorilla and chimpanzee mothers, especially first-timers, are unsure of what to do with their infant, especially if they have not learned from other females in the group showing maternal behaviour. In captivity new ape mothers have been shown film footage, or even demonstrations by a human mother and her new baby, to encourage them to breastfeed and care for the infant. In the wild other group members may crowd around curiously, unsettling the new mother so that she tries harder to withdraw from the limelight or even discards her new baby. In some cases group members are antagonistic or aggressive to the mother and her new arrival, especially if the birth has

ORANGUTAN
Average age of mother at first birth 13–14 years
Newborn average weight 1.6 (1.3–1.9) kg
Ratio mother:baby weight 22:1
Typical interbirth interval 7–8 years

GORILLA
Average age of mother at first birth 10–11 years
Newborn average weight 2.2 (1.3–3.0) kg
Ratio mother:baby weight 36:1
Typical interbirth interval 4 years

been difficult. Most human mothers go through a very different experience. The new baby, and especially its head, are proportionally large compared to her body – twice the size, in relative terms, of the gorilla infant and its mother. The nine-month human pregnancy allows the baby's brain to enlarge and develop, but not quite so much that it cannot fit through the birth canal. So the journey from the womb through the opening within the bowl-shaped pelvis (hip bone) means more effort, discomfort, pain, and time. Compared to an hour or two for other great apes, the average time for human labour and delivery is 10–15 hours for a first pregnancy, and 4–8 hours thereafter.

Birth hormones

The process of birth sees chemical upheavals in the mother's body. One of the main hormones involved is oxytocin, produced in the brain. At birth oxytocin stimulates contractions of the womb muscles, and cervical dilation so the baby can leave the womb. In the mammary glands (breasts) it stimulates the 'let down' reflex, where milk is released from the glandular parts into the milk ducts leading to the nipple (teat). Mammary glands are, of course, the hallmark of mammals, and all the apes and other primates have two. These glands are similar in situation and composition in all the great ape species, although proportionally larger and more noticeable in many human females. Oxytocin has received much attention in recent years as the so-called 'cuddle chemical' – a biologically active substance involved in affection and emotional bonding as well as the physiological processes of birth and breastfeeding. As oxytocin floods the mother's body immediately after birth, she embraces and caresses her infant, and develops the enormously powerful maternal bond with her offspring. In humans it has been shown, by measuring fluid samples, that high levels of oxytocin are associated with strong feelings of friendship, trust, and love. Experiments in other mammals, including prairie voles, show that oxytocin and another hormone, vasopressin (which affects the rate of urine formation), may be implicated in pair-bonding in adults, sexual activity, and the formation of monogamous relationships. Researchers are investigating many aspects of this interesting hormone among the great apes, such as whether some type of hypersensitivity to oxytocin is involved in the 'free love' lifestyle of bonobos. In humans oxytocin's links to sexual arousal and orgasm, to an improved short-term memory, and also to the condition of autism and its possible treatment, are all being explored.

Food on demand

Another hormone involved at birth is prolactin, made in the pituitary gland just below the brain. Its role is to stimulate the mammary tissues to enlarge and to produce breast milk. Apes and most other primates tend to be in the 'continuous contact' category of breastfeeders, where

Comparisons between the great apes

Show in particular the large size of the human newborn, both in absolute terms and relative to the mother's size, and also the orangutan female's exceptionally lengthy interval between one birth and the next – the greatest of any mammal. (Human data are averaged from typical 'natural' situations without birth control, assisted reproduction or other medical intervention.)

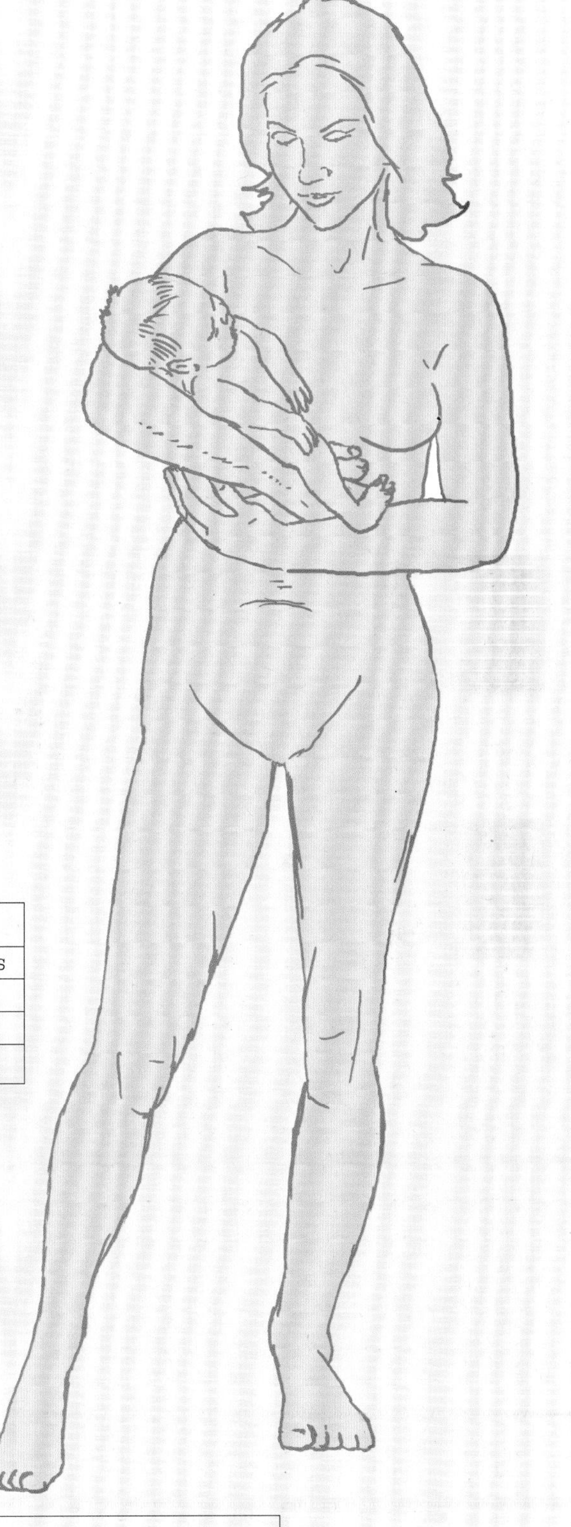

CHIMPANZEE
Average age of mother at first birth 12–15 years
Newborn average weight 1.7 kg (0.8–2.1) kg
Ratio mother:baby weight 23:1
Typical interbirth interval 5–6 years

BONOBO
Average age of mother at first birth 13–14 years
Newborn average weight 1.3 (1–1.8) kgs
Ratio mother:baby weight 24:1
Typical interbirth interval 4–5 years

offspring have access to milk for most of the time. In these species the milk is low in proteins and fats, making it relatively easy to digest, and high in carbohydrates (mainly sugars) for immediate energy. Those species that 'cache' their offspring (leave them alone while the mother goes off to feed) have high-protein, high-fat, low-carbohydrate milk, which takes longer to digest and provides more sustained nourishment. Following birth the hormones of breastfeeding take over from the hormones of pregnancy to suppress the mother's return to the menstrual cycle. This is known as lactational amenorrhea and continues until the infant is weaned, thereby preventing the mother from becoming pregnant again. Our species' received wisdom states that 'breast is best' and that breast milk is a more perfect all-round food for the new infant than formula milk. Unlike the other great apes people spend endless hours discussing the pros and cons of breast-feeding on demand as distinct from following a schedule or timetable.

HUMAN
Average age of mother at first birth 18–19 years
Newborn average weight 3.3 (2.2–4.9) kg
Ratio mother:baby weight 18:1
Typical interbirth interval 3–4 years

The Stages of Life

The helpless newborn clinging to mother. The wide-eyed innocent infant watching and listening to those around. The boisterously energetic youngster, always ready to play and learn, and the awkward adolescent in transition from carefree child to fully-fledged adult. The middle years of daily routine and social responsibility, and the onset of ageing with its mounting difficulties. All of the great apes go through all these life stages. Young great apes have the longest period of parental dependence of almost any animal, apart from perhaps elephants. The elongated life cycle and extended interbirth period are major factors contributing to the plight of all great apes – with the notable exception of our own species.

Infancy

The mother ape provides every need for her new infant – food, warmth, protection, comfort and reassurance, a means of travel, and a teacher and role model in all matters, from selecting food items to social interactions and survival instincts. The young ape's slow development, long parental dependency, and extended learning allow its brain time to soak up information on a vast scale, to understand the essentials of life, integrate into social matters, and cultivate its trademark adaptability.

The single parent family

Orangutans differ from the other great apes in that the father may well never encounter his offspring. Perhaps the only time a male chances upon a daughter or son is when they are several years old and almost ready to leave their mother – when he has returned to mate with her. Orangutans also have a more protracted period of parental dependency than the other non-human great apes. For the first two years, the infant orangutan is completely dependent on its mother. She is not just the primary carer, but the only carer it has. The mother feeds and carries her baby, and like the other non-human great apes, sleeps with it in her nest at night. In fact for the first few months mother and infant never lose physical contact. She carries it on her belly, ventral-ventral, with the infant literally clinging on for life.

At about four months the youngster starts its first tentative ventures away from its mother, although initially never more than a metre or so. Gradually it gets bolder and more inquisitive. But even by the age of one year, the growing infant still returns to maternal contact for about one-quarter of its waking time. It watches its mother feeding with great interest, noting which items she chooses. The infant has by now started to nibble its first 'solids' – often flowers from a tree that the mother visits specifically for the purpose. The infant may pluck its own food or take it directly from its mother's lips. Sometimes the mother bends branches or allows the infant to use her body as a bridge to reach the softest, tastiest morsels. An orangutan's protective instincts are so strong they can end in tragedy; she may be killed trying to prevent poachers stealing her baby for the exotic pet trade.

Baby life for gorillas

Baby great apes have a powerful grip – they need it to cling to their mother's front from their first days of life, including while travelling. Baby gorillas are quite quick to show the first signs of development. By two to three weeks they are making recognizable facial expressions, and by six to seven weeks their eyesight has developed enough for them to reach out and grasp objects. A few weeks later they can pass them to the mouth. Infant gorillas begin to crawl at about eight to ten weeks. A human baby uses the flats of its hands and its knees to crawl, and the weight of its proportionally huge head makes crawling unsteady and difficult work. A baby gorilla's relatively long arms and shorter legs make crawling easier. It is able to use its feet rather than its knees.

At four months a baby gorilla starts to take its first solid foods. The mother often helps, for example, by splitting a stem to expose the pith or stripping soft juicy leaves from a woody twig. Soon the youngster is riding on its mother's back, although she may have to help it up there. Suckling still takes place every hour or two. At six months the growing gorilla breaks maternal contact for very short periods, selects and eats its own foods, walks, and climbs by itself – and also, at this tender age,

Young at heart
(Previous page) At nine years old, this male lowland gorilla is ready to make his way in the wild, probably by leaving his natal group (the troop into which he was born). But he is not above cavorting with a two-year-old member of his extended family.

Bad hair day
Recently dry after a tropical downpour, an infant mountain gorilla takes a cautious crawl just a tiny distance away from mother.

Tender kiss
Mother of providers, a female great ape – here a mountain gorilla – cradles her new infant with unquestioning care and tenderness.

makes brave attempts to beat its (still very small) chest. After a year the youngster spends more time wandering to look for food and checking out other troop members, but never strays farther than a few metres from its mother.

A new face in the community

The birth of a chimpanzee is a time of great excitement. Others in the community are extremely interested in the new face and gather round to peer, sniff, and poke the infant. At first the mother may try to keep them away, but after a few weeks she allows trusted individuals – such as an older offspring – to 'have a go' with her baby. In return they groom her for the privilege. Touch is very important for these social great apes, and the mother-baby bond is mediated by physical contact.

Over the following weeks the young chimpanzee begins to look around with more interest and changes position on its mother more often. It may reach out to grab a leaf or twig. By six months it is taking its first walking steps and also riding on its mother's back, being too bulky now to hang from her belly. Throughout its infancy the youngster watches its mother and others intently to learn about social life. At first it eats only what its mother does – sometimes she hands over food for it to try – but as it grows older it begins to leave her occasionally to forage. Even so, it soon returns for comfort and suckling, and whimpers if it cannot find her. In fact this whimpering for mother continues for years at times of stress, through chimpanzee adolescence and even into adulthood.

Bonobo babies develop slightly more slowly than chimpanzees. They maintain physical contact with their mother for the first three months and rarely stray more than a metre away during the next three. Even after a year, infants still stay within a few metres and move less surely than chimps of the same age. Bonobos lace most activities with sex, and occasionally this extends to mother and offspring. There may be genital contact, but it is generally a form of reassurance, or to help mother ease her tension after an event such as a dispute over food.

Adult males and infants

Adult male gorillas, chimpanzees, and bonobos rarely participate directly in childcare. That said, they are usually tolerant of youngsters and put up with them crawling, clambering, and tottering around, and later even play rowdy games with them. If a male's mood turns or he becomes occupied by other matters, the mother is always on hand to rescue her infant and give it a reassuring embrace.

There are situations, however, in which male great apes turn on babies and actually kill them. Infanticide is known in orangutans but is more prominent in chimpanzees and gorillas. It is not known in bonobos. Why should a male great ape choose to kill an infant? The main reason is to improve his own chances of fatherhood. A male taking over a dominant position kills unweaned offspring who are not his. The females then stop suckling and their hormonal balance changes so that they become fertile again. The new male can then start to father his own children. This tactic ensures that his genes are propagated as quickly as possible. The alternative – waiting for infants

Holding onto mum
The infant orangutan is rarely out of physical contact with its mother for the first months of life, even as it learns to clamber and climb.

No interest in napping
Relaxed and safe in its mother's protective embrace, a western lowland gorilla infant's wide eyes and tiny ears are alert to huge amounts of incoming information.

to be weaned – could take several years. If the tendency to infanticide is genetic, it will be inherited by a male's offspring, and so these genes perpetuate as a form of sexual selection. Infanticide is known in more than 50 primate species, and is particularly common in monkeys. It is also seen in other mammals as diverse as lions and prairie dogs. In mountain gorillas it has been estimated to account for one-third of all infant deaths. It occurs when an unrelated male takes charge of a group, or when a pregnant female, or one with a young infant, is forced to transfer to another group. The threat of infanticide is a powerful factor in gorilla group cohesion. It encourages females to stay close to their silverback for the protection of themselves and their infants. The act of infanticide usually involves biting the head or groin area, and human observers come to recognize these wounds. In chimpanzees, infanticide and subsequent cannibalism have been observed not only in adult males but also adult females, being most common during times of high tension such as 'warfare' between neighbouring groups. Sometimes the mother and other adults, having tried to prevent the act, afterwards solicit a piece of the carcass to eat. Why infanticide should be unknown in bonobos is unclear. Possibly the female-dominated society has pushed aside any male instincts in this area.

Caring for another's young

Adoption of young after the death of a mother is known in various primates. Among the wild great apes however, it has only been documented in chimpanzees. Known scientifically as allomothering, the process often involves an older sibling of the orphaned infant. That sibling may be female or male. After initial resistance, the infant usually comes to make the best of the situation. By helping a close relative to survive, siblings are helping to perpetuate their shared genes (see page 200). Allomothering by a wild chimpanzee grandmother has also been observed. Cases involving females who have not borne offspring of their own for various reasons are more difficult to explain.

Childhood

In the non-human great apes the time of childhood – the juvenile stage – is one of growing independence. It includes weaning, when suckling on mother's milk ceases, and a decrease in close maternal contact such as grooming or sleeping with her in the night nest. It is also a relatively carefree time of play coupled with learning, rapid physical and mental development, and in the social apes, finding out who's who, what are acceptable forms of behaviour, and what will not be tolerated by peers and seniors.

In chimpanzees, there is a lot of evidence that a mother's behaviour traits and personality are passed on to her offspring. Chimp mothers who are generally calm, assured, confident, play with their offspring, and socialize well tend to rear offspring with similar traits. Likewise, mothers who are restless, agitated, unsure, and more withdrawn or subdued often produce youngsters who grow up with such traits.

Maternal support

Most mother great apes have considerable patience. They sit quietly as their offspring climb over them, pull their hair and ears, and get into scrapes and awkward situations. They mete out discipline with a slap or push if their youngsters go too far, but this is usually followed by reassuring hugs and grooming. As the young chimpanzee heads towards increasing independence and social interaction, from about the age of three, its mother encourages this, yet is still always around to lend comfort and reassurance by embracing or grooming. In fact some mother chimpanzees tend to groom their offspring more as they get older. This can cause tensions as the youngster strains to be off playing or exploring. But in the way of human children, if the young chimpanzee runs into trouble – a poisonous snake, an intolerant male, or a young baboon playmate who has suddenly summoned its own parent, for example – then the mother is expected to be on hand to save the situation. Gradually the mother keeps an increasing distance from her juvenile and may move away if it approaches her.

Some female chimpanzees take well to motherhood and cope from the first birth. But a few seem at a loss first time around and may even lose the baby through neglect. In general chimpanzees tend to improve their childcare with age, which is one reason why males may seek out older females in preference to younger ones (see page 210).

Young chimpanzees watch and learn all aspects of life from their mother and other adults. They study intently the different types of tool-use, such as cracking nuts on an anvil with a pebble hammer (see page 134). The juveniles may grab a piece of the food if it comes their way, and the adults tolerate this. In addition to passive observing and then imitation, which is the basic learning method, there are accounts of mothers actively helping their youngsters. One female showed her daughter the best grip on a pebble hammer, and another demonstrated to her son where to place the nut on the anvil.

How orangutans play

All great ape youngsters play. Even with the 'solitary' young orangutan, there may be an older sibling around as a social companion and playmate for the first two or three years. There is the ever-present mother too. As the youngster develops independent climbing skills from around two years of age, it begins to venture farther from her, but

Togetherness
(Previous page) In Uganda, a western lowland gorilla family spends quality time resting together in a forest clearing or baï. A female cuddles her infant as a youngster looks on, leaning against the great bulk of the group's silverback.

Ape-ing apes
Youngsters like this eastern common chimpanzee mimic or 'ape' many actions of their mothers – not just when foraging and feeding, but also to assess situations developing nearby which might spell danger.

Experiments
As young chimpanzees forage, they test out techniques such as poking and prodding with twigs, which they have gleaned by observing practised adults.

Pester practice
An infant chimpanzee tries out its upright stance, walking and arm-waving. In future months these actions could be incorporated into a full-blown fit of temper when things go wrong.

Pester power
Most human parents suffer being pestered by their offspring for toys, sweets and treats. For apes such as orangutans the main needs are a drink of milk and/or a reassuring hug.

always stays within sight. Around this time the youngster may start to hold the hand of an older sibling, its mother, or another young orangutan as it moves through the canopy, a type of movement dubbed 'buddy travel'.

Play presents a difficulty for scientific definition. In general, it is activity that seems to serve no obvious and immediate survival purpose (see page 196). However, as young animals play they improve their strength, co-ordination, decision-making, and motor skills, all of which are important for future survival. The rough-and-tumble communal play seen in young chimpanzees would be inappropriate for an orangutan high in the rainforest canopy, with few companions, and where one slip can mean death. But there are other forms of play, such as tearing up leaves, bending twigs and letting them ping back, and nonchalantly letting go of fruit to see what happens below. The young orangutan also makes practice 'play nests', copying the ones made by its mother each evening.

From the start of the juvenile period until its end, the orangutan moves from great dependence on its mother to virtual independence. By the time it is four or five years old the youngster is sleeping in its own night nest and is fully weaned. Close bonds remain, however, and the young orangutan may stay with its mother for another year or two, until after the next infant has arrived. The result of this long period of parental care is a slow reproductive rate, but a much lower infant and juvenile mortality rate than those of the other non-human great apes.

High-energy chimps

In contrast to the slightly demure nature of the orangutan's childhood, young chimpanzees and bonobos have a helter-skelter round of play, socializing with peers and adults, and generally burning energy at every opportunity. The young chimp's inventiveness at fun and games is astounding. Even when alone, these apes have a riotous time as they jump, spin, roll, somersault, leap onto rocks, and swing from branches. They also mess about with objects, throwing leaves and swiping twigs, and even tease other creatures such as baboons.

When chimpanzees get together to play it opens up even greater possibilities. Playing together is often preceded by the play-hit signal, a gesture to invite play, where the youngster raises its arm and taps or knocks potential playmates to signal that what is coming is not serious. There is also the play face – a wide grin but the top teeth covered by the upper lip. Once all is understood, the youngsters wrestle, chase, tickle, bite, and knock each other about. Almost anything is acceptable. Older siblings may join in too, sometimes rescuing the younger chimpanzees from tricky situations. All the time, like other young apes, the young chimps are extending their abilities, exploring what their bodies can do, and testing not just the world around them but also other members of their community. Play in bonobos begins at about two years of age, when the apes are still relatively unsteady and learning their movement skills. Young bonobos mostly stay within sight or earshot of their mothers until age three.

Midday fun for gorillas

Young gorillas have a somewhat more ordered social life and daily routine than chimpanzees. Playtime usually takes place around midday, while the adults are resting, but it can still be a lively affair. From two years of age gorilla youngsters spend more time with their peers and friends than with their mothers. Mutual play helps to establish their place within the small group, but is usually limited to two youngsters at a time, rather than the three or more often seen in chimpanzees. Young gorillas race about and invent all manner of activities, from 'king of the castle' on a rock or tree stump, to forward-rolling down a steep slope until they become dizzy, or collecting fruits not to eat but to throw about like balls. Adolescent males may join in, although females rarely do. As with all great apes, the adults tolerate the young bundles of fun. The silverback is supportive of his offspring and often allows them to climb over him. Sometimes a gorilla youngster 'teams up' with a junior adult male, or blackback, and follows him everywhere.

Weaning

The weaning period can be stressful for both mother and young great apes, and may coincide with the latter starting to sleep in their own nests at night. Young chimpanzees, in particular, throw temper tantrums as the mother withdraws her services. They scream, stamp, and thrash about in a fury – a sight familiar to many a human parent. The mother chimpanzee may try distraction techniques such as grooming or tickling, but she usually remains resolute. In gorillas the process is less fraught and the silverback or blackback may help to support the youngster through the process.

The age of weaning varies among individuals as well as with seasonal factors and food supply, and like several aspects of the life cycle of great apes, it tends to occur earlier in captivity. As a broad average, in the wild, weaning starts for gorillas at two and a half to three years of age, in chimpanzees at around three to four years, in orangutans at around four years and in bonobos at four to five. As the mother's milk supply dwindles, she comes back into reproductive condition. The birth of her next baby is a major signal to the juvenile that it is time to move on and become more independent.

Milestones of infancy

Paediatricians call them 'stages in acquisition of motor and cognitive skills'; most human parents know them as the 'milestones of infancy' such as sitting up, crawling, walking and talking. Broadly speaking all of the non-human apes pass through their early developmental stages in the same order, while humans differ in several major ways.

An interesting comparison of humans and the other apes, which has a major impact on the early development, concerns the rate and timing of the brain's growth. With factors such as body size taken into account, a typical human newborn is about twice as large as an average newborn baby from the other great ape species. Also it has a brain of more than twice the size, being some 13 per cent of total body weight (by adulthood this has reduced to two per cent). At no other time of life does the brain grow so quickly in relative terms as before birth, during the embryonic and fetal stages in the womb. Then in most primates, including the non-human great apes, brain growth rate sees something of a slowdown after birth. However in humans the brain's relative growth continues at a fairly high rate after birth. This has led to the idea of a '21-month pregnancy' as far as human brain development

BIRTH

The typical birth time for a mother gorilla is usually less than one hour; for other apes it may be slightly longer. From facial expressions, body movements and overall demeanour, pain features far less than in humans.

NEWBORN

Close contact and cuddling are extremely important for all newborn great apes. They cannot see very far or coordinate their movements. They instinctively cling and search for their mother's nipple.

3 MONTHS

Relatively long arms and shorter legs mean that crawling, which can start by three months in some species, is easier using the feet rather than the knees. The wrists, hands, and fingers are predisposed to knuckle-walking rather than using the flat of the hand.

4 – 5 MONTHS

Some young apes can totter unsteadily on their back legs from four or five months. But it never becomes a habitual mode of travel and usually occurs in novel situations such as play.

0 months | 1 | 2 | 3 | 4 | 5

BIRTH

The average time for labour and delivery in humans is 10-15 hours for a first baby, and 4-8 hours for subsequent offspring. Considerable pain may be endured.

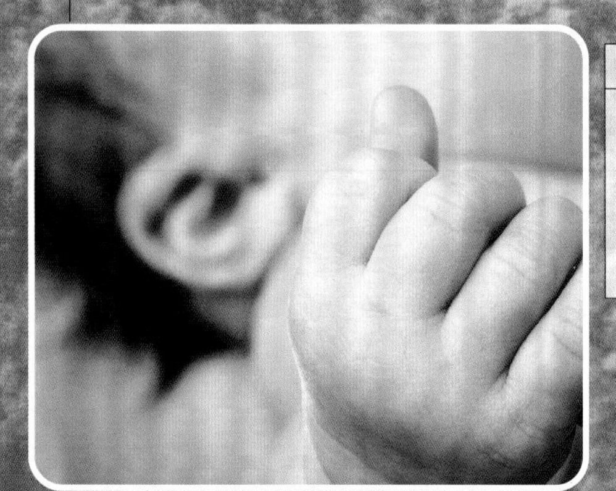

0 – 2 MONTHS

Neonatal reflexes are similar in all great apes. The grip reflex is better developed and more prolonged in the non-human species, since the baby must cling to its mother from the first day.

4 – 5 MONTHS

A baby's new-found strength at this age can be surprising for parents and baby alike. Independently rolling onto its back, performing baby 'press-ups' – (motor skills) are accompanied by a new-found sociabillity, common to all the primates at this age.

is concerned, made up of the nine months inside the womb plus the following 12 months outside. Zoologists term it 'secondary altriciality' – an altricial species produces immature new young unable to care for themselves, while precocial offspring are well developed when born or hatched, and more equipped to fend for themselves.

Importance of circumstances

As human parents know only too well, there is huge variability in the ages at which different infants go through their developmental stages. Of course this variability occurs with other ape babies too. Recent research has compared baby chimpanzees in the wild, orphaned chimpanzees brought up by humans with differing amounts of care and love, and a similar range of human babies – using observations from tragic situations in human orphanages where infants have been starved of love and attention. The results have surprised many. In these varied circumstances there is a wide amount of overlap between the two species in development of physical skills, mental and communication abilities and emotional well-being. So much so that nine-month-old chimpanzees receiving intensive levels of motherly care outscore nine-month-old humans who receive less attentive care. That hallmark physical skill of the human, bipedal walking, is welcomed by parents with a mixture of pride, elation and relief – and if it occurs earlier than the norm, perhaps a touch of haughty superiority. Other ape parents, as far as we know, do not involve themselves in the 'milestones race' or boast to others in their group about how well their offspring are doing.

6 MONTHS

Young gorillas and other apes become adept at selecting food items, grasping them and passing them to the mouth by the age of 6 months. Soon they are regularly feeding themselves.

12 MONTHS – 5 YEARS

Orangutans remain entirely dependent on their mother well into infancy. In the first year any moments apart are anxious for mother and child. Learning to climb independently occurs after 2 years.

7 – 10 MONTHS

From around 7-10 months, the human baby starts to crawl. It rests on the flats of the hands, and the comparatively long legs mean the knees are more convenient weight-bearers than the soles.

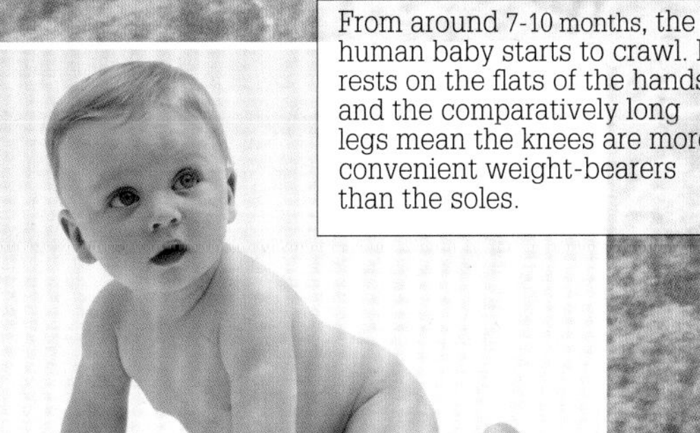

9 MONTHS

Human babies lag slightly behind other apes at feeding themselves with small items. Their hand-eye coordination is delayed, and also a parent or carer usually provides the food and even places it in the mouth.

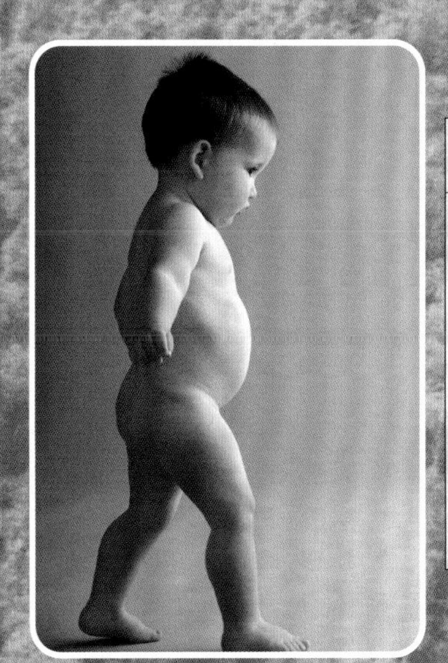

11 – 13 MONTHS

Most babies start to take faltering steps from around nine months, usually with a steadying support such as an adult hand or furniture. Unaided walking generally occurs at around 11-13 months.

Adolescence

The adolescent period sees a great ape's transition from the fun and games of childhood towards the trials of young adulthood. It includes the physical development of puberty (see page 212) and maturation of the sexual organs so that reproduction is possible, although actual parenthood may still be some years off. Some African apes have to change troops, and all attempt to rise up the social ladder. Orangutans need to establish their home ranges.

The hormones of puberty affect behaviour as well as physical growth and sexual maturation. However, apart from the male gorilla and orangutan, and the male and female human, there is no great pubertal growth spurt. There are also relatively few outward signs of the adolescent phase in non-human great apes, other than the sexual swellings in female chimpanzees and bonobos (see page 212). During puberty there is a period of female adolescent sterility, allowing females to leave their natal group and work their way into another community.

Young male chimpanzees begin their sexual careers from the age of two or three, trying to impress and even mount any interested females. Then from about seven to eight they begin to study and challenge other males and enter the adult hierarchy.

A helping hand up the social ladder

Passing through adolescence is far from an equal affair. The quality of maternal care and group dynamics both play their part. In many troops, chimpanzee mothers invest more time and effort in one or other sex of young, depending on their own circumstances within the group. A chimpanzee son is relatively high risk, since he might never become a father – but if he does, he could produce many progeny. The chimpanzee daughter is a safer bet for reproductive potential, but this is limited through her lifetime to just a few offspring. A high-status mother will support a son by delaying her next birth by a year or even two, to aid his rise through the ranks towards plenty of fathering opportunities. In this 'borrowed rank' system the son gets fast-tracked into the highly competitive male hierarchy. Lower-ranking mothers tend to invest more in daughters and give them the extra months or a year before breeding again. These patterns of reproductive behaviour are not fixed however, and differ in some chimpanzee troops.

The adolescent bonobo

Like young male chimpanzees, juvenile bonobos are sexually precocious, before becoming less active in adolescence. The adult females of their group are usually off limits, already occupied by the senior males. However adolescent male bonobos may use devious tactics to attract a female's interest. A male might approach a female resting in a tree and gain her attention by dropping fruit or branches before attempting a quick copulation, all done on the quiet, out of direct sight of dominant males.

Maturing female bonobos begin to break their maternal bonds at six to seven years, eventually leaving for a new community. The males tend to remain close to their mothers throughout adulthood, and as in chimpanzees, the mother's status determines her son's rank as he enters adulthood.

Studied indifference

The adolescent great ape is much the same in all species (although in non-human apes adolescence is complete by 12 years). Torn between acting like a child, playing games and having fun, and behaving like an adult, he or she moves towards feigning indifference and studied reflection. Spending time alone, the 'teenager' can be listless, tapping branches, swatting flies, and indulging in other minor activities.

Gorilla females reach maturity from the age of about six or seven to nine or ten. As mentioned earlier, the female then goes through a period of 'adolescent sterility'. As the young male or blackback gorilla matures (from about eight or nine to twelve years of age) his fate depends partly on the age and condition of the silverback in his group (see page 184).

Orangutan adolescence lasts from about five years of age to eight, sometimes beyond. During this time, although still in contact with their mothers, adolescents may form temporary parties with their peers to travel or feed. Females start to show an interest in the local dominant male as they try to establish their own home ranges near those of their mothers. Adolescent males leave home, dispersing farther, and enter the sexually mature but socially junior 'unflanged' phase, as described on page 219.

Hanging out
Adolescent bonobos play it cool as they take a break from group life. Or should that be 'grope life', since these apes develop sexual tendencies from an early age.

Adulthood

It's soon after the six o'clock sunrise. Time for gorillas to rise gently, stretch, yawn and scratch, urinate and defecate, check all is well, and reach out to grab something to munch on. Next comes more food, and more, while moving as little as possible from one feeding position to another. Chewing all morning is tiring, so there's a midday siesta for a couple of hours to rest and digest. This is the time for some self-grooming and perhaps socializing and mutual grooming with a friend. A few youngsters disturb the peace as they frolic around, but that's their way, and after all, they are the future.

Rousing from the midday break, the group moves on. They stop for a few minutes here and there for an occasional snack. The destination is another patch of forest where tomorrow's food is ready and waiting. It's now past five o'clock and the tropical dusk approaches. Not much has happened today. Bedtime is making a nest, a yawn, and settling down to sleep. Of course not every gorilla day is like this. There may be social tensions, dominance issues to settle, the occasional illness, and perhaps a neighbouring gorilla group to negotiate. When foraging, the group members tend to spread out in case of squabbles over a particularly tasty patch. When moving to new areas the silverback, a senior female or a blackback-in-waiting leads the way, often in single file, while another of these brings up the rear. The distance covered each day varies with season, group size, local population density, and other factors. Generally it is from less than 500 metres (1,600 feet) for some mountain gorillas to 3km (1.85 miles) or more for lowland gorillas.

A life on the go

Chimpanzee adults have busier days with less routine than gorillas. They spend about half of the daylight hours foraging and feeding in trees and on the ground, and travelling between food sources. The rest of the time is largely taken up with social activity. For the males there are alliances to maintain and for mothers, youngsters to feed and watch over. Senior males may be especially active, patrolling the community's peripheral range if they sense a threat from neighbours, or setting up hunting parties. At night, like all the non-human great apes, chimpanzees settle to sleep in self-made nests, invariably sleeping alone, apart from mothers with their offspring.

Approximately two-fifths of the average bonobo day is spent resting, grooming, and socializing, another two-fifths searching for food and eating, usually in the branches at heights of 20 to 40 metres (65 to 130 feet), and most of the rest travelling between stopovers. Bonobos cover distances of around 2 km (over a mile) a day and are more gregarious than chimpanzees, especially at night. Towards the end of day party members and the community call to each other and gather to nest.

Regular habits

In the early morning of a typical day orangutans listen to the sounds of many creatures, including birds, monkeys, and gibbons. Special attention is paid to the penetrating 'long calls' of dominant male orangutans, which could signify intent to mate. Grooming, defecation and sniffing the air are all important activities. The priority, however, is breakfast. Like the other great apes, orangutans use various clues to find food. These include sights, sounds, and smells – the powerful odour of durians, the sight of colourful ripe fruits, the movements of other creatures such as monkeys and birds towards food sources, and the feeding noises of other animals. Once breakfast has been found other orangutans might arrive at the same tree, but if the fruits are plentiful, there is no aggression and the various apes feed in different parts of the tree.

Orangutan forests have several sets of fruiting patterns (see page 151). Many trees follow a 'masting' system, with very few seeds over a few years, then plentiful crops in a single season across the forest. This type of pattern is linked to fluctuations in rainfall, which are in turn influenced by global climate cycles, especially El Niño events. During these times of plenty, orangutans fill up on food and accumulate reserves of body fat.

The midday nap

Self-grooming is another important daily activity for orangutans, to remove bits of leaves, fruits, and pests from the skin and fur. There are regular pauses in this process to rest, listen, and search for bromeliad flowers that have trapped fresh water in the bowl shape of their leaves. Living up in the canopy, water is sipped from many sources. Bromeliads are also eaten.

Around late morning orangutans forgo feeding and settle down to rest, perhaps building a day nest – a simpler version of a night one. There is further grooming, especially by mothers of infants, always with one eye open for dangers such as pythons. The afternoon is occupied with more foraging, sometimes guided by hornbills and monkeys to fruiting trees or strangler figs. Where several orangutans have formed temporary parties to travel or feed, there are social assessments to make (see page 172). As the light wanes it's time to search around and assess a few trees to find a suitable night nest site. In the course of one day, orangutan travelling distances range from just a few metres to several kilometres, depending on the availability of fruits and position in the home range.

Midday stretch
As noon approaches a drowsy silverback mountain gorilla extends and yawns, belly nicely full , ready to take a break from the tiring time of the morning feed.

Watchful eye
Chimpanzees are ever alert to the social dynamics of their group and who is cosying up to whom, which means truly restful periods are rare.

A place to sleep

Nests are an important part of every great ape's life. Bonobos, chimpanzees, and orangutans make their nests in trees, male gorillas on the ground. We humans have armchairs and beds. Nests provide comfort, allow restful sleep, give some protection against predators and poisonous creepy crawlies, and lift the body clear of the damp ground. It takes a few years to master nest-building, and young great apes make practice nests as part of their play. The only time wild great apes share their nests is when mothers curl up with their youngsters.

There are many similarities between the wild great apes in nest building behaviour. The first task is to carefully select a site, such as a fork in strong branches or a dense crown of smaller boughs. A few branches are then bent over and down with hands and feet, and interlocked or jammed together, making a leafy platform. Broken-off branches may be interwoven for greater stability and security. A lining of leaves and softer materials completes the job. A hurried chimp can knock up a nest in a minute or two, while a relaxed orangutan might take as long as 20 minutes. Day nests are simpler than night ones.

Behaviour at the nest

Some orangutans make a 'raspberry' sound during or after nest-building, which differs in various regions. These red apes are also known to re-use an old nest occasionally, giving it a fresh lining each time. The nest might be covered with leafy branches or a single large leaf as a sort of roof or umbrella. During a baby orangutan's first months its mother avoids building her nests too near to heavily fruiting trees, because other fruit-eaters collect there and in turn might attract predators. Some orangutans have the habit of decorating or lining their nests with brightly coloured objects like green leafy twigs.

Big male gorillas usually sleep on ground. They make nests after a fashion by bending and twisting stems of plants such as bamboo and trampling the vegetation to shape a 'sprung mattress' to keep them off the damp ground. Defecation occurs in or near gorilla nests, so counting nests and their dung helps researchers to estimate the number and size of gorillas in a group. Mountain gorillas make more ground nests than other gorillas, as there are fewer tall trees in their habitat. Young gorillas practise the art of nest-building from about nine months old. Some youngsters are able to make their own nests to sleep in by the time they are three. Up to 10 chimpanzees may nest in one large tree, reflecting the composition of their last travelling party. In some areas chimps favour particular trees such as oil palm, muhimbi (Ugandan ironwood) or sassy. On average nests are built between 5 and 20 metres (16 to 65 feet) from the ground. Bonobos usually nest 15 to 30 metres (50 to 100 feet) up and are more socially organized than chimpanzees, with the chief females nesting near each other and higher up than the males. Day and night nests fulfil several purposes apart from sleep, including play and grooming. Some bonobos gather leaves and twigs for their nests from several different trees – a trait not seen in chimpanzees or any of the other great apes.

Ready to retire
As the chief male emphasizes his rule, a family group of mountain gorillas prepares for the night at Mgahinga National Park, Uganda. He will stay ground-based while the others choose nesting trees.

Unnatural loss
For creatures so well adapted to their environments, ape accidental injuries are rare. This mountain gorilla's stumpy fingers result from a lucky escape out of a poacher's snare.

Illness, injury, and death

The close relationships between the great apes, including ourselves, mean that we all suffer a phalanx of similar infections and other diseases including growths, metabolic disruptions, malformations, degenerations and other health problems.

Chimpanzees and bonobos, and to a lesser extent gorillas, share so many genes with us that it is not surprising they are vulnerable to many of the viruses, bacteria, and other microbes which afflict humans. The long list includes the infamous Ebola virus, the common cold, influenza, pneumonia, whooping cough, tuberculosis, measles, yellow fever, HIV-AIDS, poliomyelitis, chickenpox, herpes, hepatitis, staphylococcus and streptococcus. The threats posed to both wild great apes and to the humans who come near to them – from field workers to ecotourists and bushmeat hunters – are exceptionally grave and discussed on pages 262–263.

Chimpanzees and HIV

Even a mild respiratory infection can quickly kill a young chimpanzee. At the other end of the scale, two types of human immunodeficiency virus, HIV-1 and HIV-2, are known to affect people and be derived from other primates. HIVs are regarded as strains of SIVs, Simian Immunodeficiency Viruses, the overall name for viruses of this type. HIV-1 is the most common and widespread – HIV-2 is restricted mainly to West Africa. Studies of HIV-1's structure, and the immune defence systems of ourselves and chimpanzees, suggest that it evolved from a strain of SIV known as SIVcpz, which is found in central chimpanzees. This virus mutated and 'jumped' between species some time in the last century. In all likelihood the method by which this occurred was the eating of infected chimpanzee flesh as bushmeat. Molecular studies of the virus suggest that this probably happened in West Africa. (HIV-2 is probably derived from another strain of SIV, SIV-sm, which is found in the monkeys known as sooty mangabeys.)

Chimpanzees infected with SIVcpz do not suffer destruction of the immune system or the secondary infections and cancers that are the AIDS-related problems that humans face with HIV-1. Chimpanzees injected with HIV as part of medical research have shown themselves to be similarly unaffected, showing no symptoms at all.

As well as viruses, wild great apes contract many of the parasites that humans do in their habitats, such as schistosomiasis (bilharzia or snail fever), filariasis (which includes elephantiasis), the digestive disturbances of giardiasis and salmonella, and fungal infestations such as ringworm (tinea). Mountain gorillas have also been known to suffer outbreaks of scabies or sarcoptic mange, caused by a tiny mite (related to spiders). Orangutans can become infected by many of the diseases mentioned above, as well as cholera, malaria, various types of hepatitis, and leptospirosis (Weil's disease).

Injuries from accidents

In many parts of the world the human body's main threat of injury is a road accident. The equivalent for the other great apes is a fall. This is nowhere near as common as we might imagine considering their lifestyles, but it does occur, usually due to a bough breaking, and less commonly from a slipped grip. Human observers have noted that falls tend to occur with apes – especially orangutans and chimpanzees – who are already suffering from other problems, such as an infection, malnutrition, or some kind of digestive upset, such as food poisoning. The riskiest times are youth, when bold, inquisitive behaviour often leads inexperienced youngsters to take chances, and of course old age.

Violence and murder

Around the world, humans in many big cities sometimes find themselves in situations involving dangerous rivalry – 'turf' disputes for drug dealing, gang culture, and the like. The other great apes do not have guns or knives but they do have powerful teeth, and their canines

are especially dangerous. Mature male chimpanzees and gorillas in particular are not afraid to use these weapons on rivals or opponents. Several instances have been recorded through the years of collaborative attacks by chimpanzee hunting parties that have wounded neighbours or intruders from other chimpanzee communities so badly that they have died as a result – and sometimes been eaten.

Silverback gorilla contests can extend beyond the ritualized stage, which is mostly bluff with little serious contact, to real scraps with savage bites and scratches. Occasionally these can end in death. At the Lopé National Park in Gabon, researchers were able to use DNA samples collected 'non-invasively' (from hair roots) to solve a 'murder mystery'. A silverback male known as Porthos had been found dead, killed by severe canine wounds to his body. Following his death, the group he had ruled over disbanded. Using the genetic evidence they had collected, the researchers identified the probable culprit as a nearby male called Yorick. It is known that a female from Porthos' group moved to Yorick's group the day after the 'murder', thereby providing some interesting side-information on female inter-group transfers. Before the genetic evidence was collected and analysed this mystery was something of a 'cold case'. Porthos died in 1993 but genetic science did not catch up with his assassin until 2007.

Other work at Lopé and at Taï Forest has shown that not all bite marks sustained by great apes are from same-species rivals. Of all the predators in Africa, leopards are probably the biggest scourge of the great apes. Whether these big cats purposefully track chimpanzees and gorillas on a regular basis, or whether a few individual leopards have taken to hunting this type of prey, or, indeed, whether the kills are chance encounters, remains unclear (see page 252).

Old age and death

The oldest known non-human primate (allegedly) is perhaps the most famous, Cheeta. (Cheeta was his stage name – his 'real' name is Jiggs.) Cheeta starred as Tarzan's sidekick in several Hollywood movies during the 1930s and 1940s. His career continued through to the 1960s and culminated in the epic *Doctor Doolitle* in 1967. It is thought that Cheeta was born in Liberia in about 1932, so he had a special 70th party in 2002. He has been suggested for an Oscar nomination several times and now passes his time painting in the 'ape-stract' style.

The authenticity of Cheeta's age has been brought into doubt several times. Some people have questioned whether a chimpanzee could really make it to such a great age, even with all modern comforts of life around him. Scientific proof of his age is actually impossible to obtain.

Despite the doubts surrounding Cheeta's story, there is no doubt that the great apes are certainly long-lived. Among mammals longevity is broadly related to body size – elephants can achieve 80 or perhaps even 90 years of age, while some whales are thought to live well beyond 100. Many humans have attained life spans of more than 100 years, of course. Some have made it to 120. Non-human great ape life spans are tricky to estimate in the wild, and life in captivity does not really replicate the natural situation. Several estimates combine to show that a 'good maximum'for chimpanzees and bonobos is around 40–50 years, and possibly 60 years, especially in captivity; estimates are similar for orangutans.

Gorillas, despite their larger size, reach a 'good old age' at around 40–50 years. Both captive and wild data suggest that for chimpanzees, females have greater longevity than males, while a long-term study of wild Sumatran orangutans shows the reverse. In early 2009 Gregoire, believed to be the oldest chimpanzee in Africa, passed away peacefully in his sleep at Tchimpounga Chimpanzee Rehabilitation Centre, Democratic Republic of Congo. He had spent 40 bleak years in a bare cage at Brazzaville Zoo before being rescued by Jane Goodall Institute staff in 1997, when he 'came back to life' under a new regime of healthy diet, care, and attention. His estimated age at passing was 66 years. A few months before this, Jenny, the western lowland gorilla, had passed away at Dallas Zoo, home for almost all of her life, at the ripe old age of 55 years.

Mourning

All great apes are known to mourn those close to them upon death. This is especially poignant for mothers with their young. Human observers – even seasoned field workers who have sampled most of the drama that the wild ape world has to offer – find it hard not to weep. Gorilla and chimpanzee mothers have been seen to hold their newly deceased babies, stroke and caress them, gently prod and shake them for signs of life, and place them on their backs to carry them around as in life. This can go on for days, even weeks. Members of a gorilla group have also been known to stay with the dead body of one of their members, disrupting their usual routine of moving on in search of food. Sometimes they may even place twigs and leaves over the remains.

In 1989 at the Taï Forest site, a chimpanzee called Tina was attacked, killed and mutilated by a leopard. Several members of her group soon gathered around the corpse, calling loudly. Then the group fell silent. Occasionally one individual would approach the corpse and touch it or pull the foot, as if to wake it. Then two high-ranking males groomed Tina's dead body for about an hour. The group stayed with the corpse for six hours before moving on. Over the next few days they returned to it as if standing guard, before the body was finally taken by a scavenger.

Signs of ageing
Greying hair, flabby flesh, drooping features, wrinkles, duller senses, slowed reactions – it's a familiar ageing story for all the great apes, like this very mature orangutan.

Growing Threats

All the wild great apes are listed by the IUCN as either Endangered or Critically Endangered: this means that, at best, they have a very high risk of extinction in the wild in the near future. The statistics surrounding their declining numbers are frightening, and low birth rates mean that their populations are slow to recover from any setback. To make matters worse, most of the setbacks are the result of human greed and human carelessness: habitat destruction, transmission of disease, uncontrolled hunting and the depredations of war are all taking their toll on our nearest relatives.

Natural threats

How natural causes take their toll

Silent stalk
The clouded leopard, the most arboreal (tree-dwelling) of the 'big' cats, is an occasional orangutan killer – and is also itself at great risk of habitat loss and persecution.

Human activity has caused widespread devastation among wild ape populations in both Africa and Asia. Nature, however, can also be harsh. Fires, floods, and droughts not only kill wild animals, they cause habitat loss and food shortages, and generally make life difficult for any survivors. Disturbances of this sort can also upset the delicate balance between predators and prey.

Global warming

In 2007, a study was carried out to assess the impact of climate change on orangutan populations in Indonesia. The results were alarming. In times of drought, orangutans head towards rivers and this brings them into conflict with human beings, who tend to do the same thing. Occasional droughts are not unheard of in the forests orangutans inhabit, but in recent decades these droughts have been longer and more extreme. During the drought and associated forest fires of 1997–98 (see pages 258–9) a third of the wild orangutan population perished from starvation, burning, or at the hands of people whose gardens they invaded. Further serious fires in 2006 killed an estimated 1,000 orangutans, a figure all the more worrying because many were females with young, being less able than adult males to flee the flames. The loss of so many females of breeding age has serious long-term implications for many populations' chances of recovery.

As if the effect of droughts and fires on the apes themselves was not enough, now scientists have started to become concerned that climate change will affect the plants in the forests of Borneo, including many species that produce fruit on which Bornean orangutans feed. If food becomes short, this will have a knock-on effect on the apes' movements and ultimately on their breeding patterns. There is also a possibility that climate change may result in an increase in the occurrence of outbreaks of malaria.

Predation

Adult wild great apes have few natural predators. The most widespread in Africa is the leopard. Christophe Boesch, who has been involved in chimpanzee studies in Taï National Park in Côte d'Ivoire for 30 years, believes that leopards are responsible for more chimpanzee deaths than any other factor, and that the population density of leopards in a given area may influence the size of chimpanzee groups. This ability to change group size in response to a perceived threat is another example of chimpanzee adaptability. In the forest habitat of Taï, chimpanzees live in larger parties when food is abundant and predators are scarce – similar grouping patterns have been observed among bonobos in similarly dense habitats. It is thought that the ability of smaller parties to move more quietly may help prevent leopards from locating chimpanzees or bonobos and launching an attack. In the more open habitats of Gombe and Mahale in Tanzania, leopards are fewer and chimpanzees live in groups that are relatively large. The cats don't have it all their own way, however. In an experiment in Gabon that used a mechanical leopard, chimpanzees ganged up against the

The loss of so many females of breeding age has long-term implications for the apes' recovery.

Double trouble?
(Previous page) Ecotourism can bring much needed income to local people for conservation. But risks include disease transmission, altering the apes' behaviour in unnatural ways, and setting up a system of financial support subject to political instability and global recession.

Flood risk
More extreme weather due to climate change, coupled with logged forests, are a recipe for flash floods.

predator and used sticks and stones to attack it. Gorillas are known to run away from leopards and researchers have recorded signs of the gorilla 'fear odour' after an encounter between the two. However, although there has been the occasional suspected case of a leopard killing a gorilla, the adult gorillas' size and strength are normally sufficient to protect both themselves and their young from attack. The Bornean clouded leopard of Borneo and Sumatra (recently designated as a separate species from its Asian mainland counterpart) spends some time on the ground but also hunts regularly in trees. It has been known to take juvenile orangutans and also snatch an infant when the mother is distracted. Large pythons, black eagles and other sizeable birds of prey are also known to be a threat.

Replaceable baby
In stressful times, especially when the leadership of a gorilla group changes, adults tussle over an infant – who might not survive.

Surviving childhood

The biggest challenge to any wild creature is surviving infancy. Orangutans have got this down to a fine art: they have the longest 'childhood' of any animal other than humans, with mothers caring for their offspring for up to eight years. As a result of this care, infant mortality from natural causes is almost unheard of. About 90 per cent of youngsters make it through the period when they are dependent on their mother, as opposed to just 50 per cent of chimpanzees and as few as 35 per cent in some groups of lowland gorillas.

Infanticide

One of the principal causes of infant mortality among many wild animals is death at the hands of an adult male. The loss of an infant brings the females of many species back into reproductive condition, so infanticide is actually an advantageous behaviour for a male who is not the father: it gives him the opportunity to mate and pass on his own genes. This is particularly significant among the wild great apes, where, with the exception of the bonobo, the mother is sexually unavailable throughout the period when her offspring is dependent on her.

Among orangutans, non-dominant males frequently force their sexual attentions on females and father perhaps 50 per cent of all orangutan offspring. In chimpanzee communities, females voluntarily have sex with both dominant and subordinate males. In both chimpanzees and orangutans a dominant male will kill the unweaned offspring of lesser rivals.

Most gorillas live in harem-style groups, with a number of females and their offspring presided over by a single mature silverback male. Females often transfer between groups, perhaps to seek a more suitable mate or to move into a territory with less competition for food. A group may also disperse when a silverback dies. Any infant belonging to a female who moves from one group to another is at risk from the new silverback. In mountain gorillas, infanticide accounts for 37 per cent of deaths of youngsters under the age of three.

Some gorillas, however, live in multimale groups in which the males are probably related. In this case, when the silverback dies, another male from the group usually takes over and infanticide is not a risk (the youngsters in the group are likely to be nieces and nephews of the new silverback, so are already carrying a number of his genes). With bonobos, where everyone mates with everyone else, no one knows which males have fathered which young. This fact, together with strong bonding between the females, helps protect the young. Infanticide is unknown among bonobos, and the survival rate for youngsters is correspondingly higher than among chimpanzees or gorillas.

Conflict between adults

Chimpanzees from neighbouring groups will kill each other to protect their territory and their females from intrusion by rival males. Gorillas also fight amongst themselves – 62 per cent of injuries to adult gorillas in the Virunga Volcanoes area have been shown to be the result of confrontations between groups. Fights between gorilla groups occasionally leave fatalities. Adult male orangutans will also fight with one another – sometimes to the death – but for the most part they prefer simply to avoid each other altogether. Bonobos are generally much less aggressive than other wild great apes and confrontations are rare.

Disease

A study of the chimpanzees in Mahale Mountains National Park in Tanzania found that disease was the principal cause of death. Some of the diseases chimpanzees there died from were likely to have been the result of human contact: as many as 12 are thought to have died from a flu-like infection in a two-month period in 2006. As well as infectious diseases, both chimpanzees and gorillas also suffer from intestinal parasites, skin lice, and sometimes yaws, which causes a necrosis (death of tissue) similar to leprosy.

The most common cause of death among mountain gorillas is pneumonia. These wild great apes also suffer from colds, chills, laryngitis, and other similar diseases brought on by the cold, damp climate in which they live. Older individuals eventually lose their teeth and may starve to death because they are unable to chew their food properly.

According to one study, about 50 per cent of wild orangutans carry a malaria parasite, although they do not seem to become ill as a result of this infection. Dysentery and melioidosis, a potentially fatal bacterial disease, are also widespread among orangutans.

> Most gorillas live in harem-style groups, with a number of females and their offspring presided over by a single mature silverback male.

Old age

In the wild, all of the great apes can live beyond 50 years. Bonobos and orangutans may have a life expectancy of 45–55 years, with chimpanzees perhaps slightly less, and gorillas 40–50 years.

Habitat destruction

Apes' needs fall victim to human demands

The single greatest threat facing the wild great apes today is habitat destruction. They are all largely creatures of the forest, requiring extensive areas of wilderness in order to survive. The rate at which their homes are disappearing is startling to say the least. Between 1985 and 1997, the forests of Sumatra were reduced by 61 per cent, for example. Across the whole of Indonesia deforestation has averaged 2 million hectares per year since 1996 – that means an area the size of the Netherlands is cleared every two years.

Nor are protected areas immune: 56 per cent of the protected lowland forest of Kalimantan (the Indonesian part of the island of Borneo) was cut down between 1985 and 2001. There are two main reasons for all of the habitat destruction that has happened and continues to happen in Indonesia: logging and palm-oil plantations. All of these operations, legal and otherwise, of course need infrastructure: access roads, logging camps, and means of shipping out their produce. In the peat-swamp forests of Kalimantan there have even been canals built specifically to transport illegally felled logs; these canals not only contribute to an illegal, unsustainable trade but also dehydrate the forests.

Domestic demand for Indonesian timber far exceeds what can be supplied by legal, sustainable means.

Logging

Some logging is legal. In Indonesia, an estimated 73–88 per cent is not. In the late 1990s the Indonesian government announced that legal logging had been reduced, with supplies from natural forest, declining from 17 million cubic metres to 8 million cubic metres over a five-year period. During that time, however, the total amount of timber taken from Indonesian forests was estimated at 70-80 million cubic metres.

There are a number of economic reasons for this disparity. Domestic demand for timber in Indonesia far exceeds what can be supplied by legal, sustainable means. So does the capacity of Indonesia's 1,600 timber mills. This means that if licensed operators working on a timber concession harvest more wood than their licence permits, or if they move into land outside their concession (which may often be protected land, as these areas are among the few that still have viable quantities of valuable timber), they will always be able to find a ready market for their harvest.

Nor can Indonesian operators take all of the blame – many of the local mills are controlled by multinational companies. Some raw wood and wood products are smuggled out of the country, others are processed as if from legal operations, and all soon find their way on to an international market. Western demand fuels much of this illegal trade – because shipments change hands frequently before reaching the end user and are effectively 'laundered' of their Indonesian connections, it is easy to plead innocence or turn a blind eye.

Palm-oil production

Few people in the West have heard of palm oil, yet we consume it here in massive quantities. Palm oil is extracted from the fruit and seeds of

Handy product Western consumption encourages burgeoning oil-palm-seed production in South-east Asia, fuelling local economies but devastating natural habitats.

Easy clearance After the large valuable trees are cleared for timber, burning prepares the way for local agriculture.

the oil palm (*Elaeis guineensis*) and is used in everything from biofuel to food. It is cheap, stable at high temperatures, and provides a healthy alternative to the trans fats associated with heart disease. It is found in 10 per cent of supermarket products, from margarine to lipsticks. Between them Indonesia and Malaysia produce 03 per cent of the world's palm oil, and it is estimated that in Indonesia alone an additional 30,000 square km (11,600 square miles) of land will be given over to palm-oil production every year until 2020.

To make matters worse, palm-oil plantations tend to be located on newly cleared forest land, despite the fact that Indonesia and Malaysia have no shortage of abandoned agricultural land. Again, economics provide the explanation: oil palms do not become productive until the trees are at least three years old. Selling the timber from cleared forest helps finance the operation in its early years.

Local income
Logging near Kinshasa, on the River Congo, continues apace with few local controls. From one viewpoint, it does feed many families.

Burning bush

The easiest way to clear a forest – whether a small area to create a road or campsite, or a larger area in which to plant crops – is to burn it. Illegal fires can quickly get out of hand in dry areas, and in the peat-swamp forests, where the peat burns as well as the trees, they can smoulder underground for long periods. In El Niño years – when changes in ocean and atmospheric conditions make Borneo unusually dry – fires are most likely to spread. In 1997–98, the worst El Niño period on record, a million square km (almost 400,000 square miles) of land on Borneo and Sumatra were burnt and 95 per cent of the forest of Kutai National Park in Kalimantan was destroyed. In the process some 2.67 billion tonnes of carbon dioxide were released into the atmosphere.

Less forest can support fewer animals. What habitat remains tends to be in isolated pockets, which restricts the orangutan's range.

The figures for 2005–06, another El Niño period, were less devastating, but they resulted in heated mutual criticism between the governments of Indonesia, Singapore, and Malaysia. While most of the fires occurred within Indonesia's borders, the Indonesian government blamed Malaysian timber firms for their involvement in the exploitation of Indonesian forests. So how has all of this affected wild orangutans? Well, clearly a shrinking natural habitat will only be able to support a smaller number of animals. On top of that, what habitat does remain tends to be in isolated pockets, which restricts the orangutans' range. Then there is the obvious effect of fire. Animals may be trapped and burned to death. Statistics concerning forest fires between 2000 and 2005 show that in three of the four most recent years, fires in conservation or protected areas (where orangutan populations are most concentrated) accounted for anything from 56 to 95 per cent of all orangutan deaths. Fire and other forms of forest clearance can also affect orangutans that survive their immediate impact. With their old homes destroyed, these animals are forced to adapt or move on. Reports from Kalimantan, where oil-palm plantations have burgeoned in recent years, tell of orangutans wandering into areas that used to form part of their range but that are now covered in palm trees. Confused and hungry, they have tried to feed on the young trees and been shot by plantation workers who were instructed to protect their crop, or who were frightened and wanted to protect themselves. Some 120 Bornean orangutans were taken into rescue centres in central Kalimantan in 2006, all suffering from either the effects of fire or wounds inflicted on them by villagers. Trauma is another effect of habitat destruction that cannot be ignored: orangutans driven from their home ranges are forced closer together into smaller, less familiar, and more crowded areas. The resultant stress leads to reduced breeding success, with fewer babies being born and more youngsters dying.

The resultant stress leads to reduced breeding success: fewer babies being born and more youngsters dying.

The combination of all these factors has already had a massive impact on orangutan numbers. It is estimated that a third of the entire Bornean orangutan population was destroyed by the fires of 1997–98 alone.

The advance into Africa

Sumatra and Borneo are not the only ape homelands where habitat destruction is a cause for concern. Statistics concerning the Democratic Republic of Congo are unreliable but those thought to be the most accurate suggest that timber concessions may cover as much of 55 per cent of the bonobo's range – even the official calculation by the government of the country puts the figure at 24 per cent. For a species that occurs nowhere else in the world, and has a total population of less than 20,000 individuals, this is a lot of habitat to jeopardize.

In Gabon, although some 11 per cent of all land has been given over to national parks, all of which contain at least one great ape species, legal logging takes place in half of the forests (and some 'legal' prospecting for oil in at least one of its national parks). In Cameroon, satellite images show logging roads in the least accessible parts of the country, potentially fragmenting the habitat of gorillas. Chimpanzees, which also live in the area, are more adaptable creatures and less likely to be affected in the same way. As for orangutans, this sort of fragmentation can cause stress, which may lead to less successful breeding.

A vicious circle

Falling numbers of wild apes can have an impact on other forest species too. Apes are important dispersers of many types of seed, for example. More than half of the bonobo's diet is made up of fruit, and they feed on a wide variety of species. Bonobos are known to eat the fruits of more than 60 different rainforest trees, and the true total number may be much higher. As bonobos have large home ranges and regularly travel about 2 km (more than 1 mile) a day – often transporting their food long distances before eating it, and then moving on elsewhere to sleep – they are a very important link in the chain that ensures the forest's existence. The seeds of one tree, *Pancovia laurentii*, which is related to the maples and horse chestnuts, cannot germinate without first passing through a bonobo's gut. And the fruit of the liana *Carpodinus gentilii* is too hard for most monkeys and other smaller primates to open and relies largely on the bonobo for its dispersal.

Chimpanzees are even more fruity than bonobos: fruit may comprise more than 90 per cent of their diet. Some of the seeds they ingest germinate very poorly if they simply fall from the parent tree, but do well if carried farther away. Gorillas also play their part in seed dispersal. In Lopé National Park in Gabon, where gorillas travel around 2–3km (1–2 miles) a day, they are the sole dispersers of most of the dominant tree species.

Hunting and poaching

Ape vs naked ape

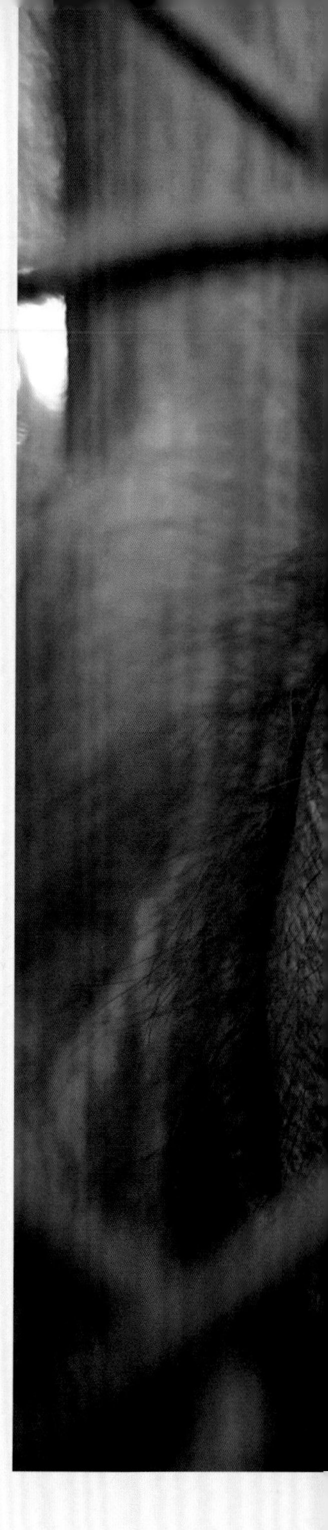

For centuries some indigenous peoples have eaten the meat of great apes; they have ground up body parts to make ingredients for medicine, fashioned bones into protective amulets, and even, in the case of the Dayaks of Borneo, used orangutan skulls believing they could impart spiritual energy. Others revered the apes as a reincarnation of their ancestors, while some refrained from eating their meat because it is forbidden under Islamic law.

As local taboos become less powerful as a result of the break-up of traditional communities, and as growing human populations demand increasing amounts of food, ingrained reverence for great apes has begun to give way to the need for meat and the prospect of financial gain.

Legal and illegal trade

Incredible as it may seem, there are still countries where great apes live wild that defy international law and CITES bans, and license the killing of these protected species. As recently as 2005, the minister in charge of tourism and the environment in the Democratic Republic of Congo allowed a mountain gorilla to be shot for a fee of US$500 and a live one to be captured for US$1000. The payment of 'an appropriate fee' also permitted the payer to catch or shoot bonobos and chimpanzees, despite the fact that the Congo was then, and still is, a (voluntary) Party to the CITES agreements. Since the early part of the 20th century, there has been a constant demand for young apes from the wild. They have been captured as pets, for private collections and zoos, to feature in films, as attractions in amusement parks and bars, and to be used in medical research. This trade is now illegal, but it persists in a lucrative black-market form: it is said that an orangutan offered for sale in the late 1990s could have fetched up to US$20,000. The trade in young apes concentrates mainly on chimpanzees and orangutans, for the simple reason that baby gorillas taken from their mothers usually die before they reach their destination (though even with chimpanzees and orangutans the survival rate can be as low as one in five). The babies taken are not the only apes harmed by this trade: adults fight fiercely to defend their young and often lose their lives in the process. At one point during the 1996–2002 war in the Democratic Republic of Congo, 12 infant bonobos were offered for sale in Kinshasa market. It was estimated that in the process of catching these babies, between 60 and 120 adults had been killed. Their bodies were butchered for meat.

Bushmeat, always eaten by the poor, has latterly become a delicacy and a status symbol among the African elite.

Bushmeat

The illegal bushmeat trade is vast – an estimated US$1 billion dollars a year – and although the wild great apes form only a small proportion of this, some estimates say that several thousand apes are killed each year in central and western Africa. Bushmeat, in the past always eaten by the poor, has latterly become a delicacy and a status symbol among the African elite. It has also become a serious business for many hunters. A butchered chimpanzee or gorilla carcass is now worth between US$20 and US$25 in central Africa – nearly a month's wages for the average person in Mali and almost three months' in the Democratic Republic of Congo. Those parts that cannot be eaten may be turned to other uses. A gorilla's hand severed from the carcass might be sold to make a souvenir ashtray, while the skin and head could adorn someone's trophy-room wall. If the current impact of hunting for bushmeat continues, it alone will be sufficient to render

March of progress? (Previous page) Near Miri in Sarawak, Borneo, the orangutans' natural forest (left) is logged for hardwood timbers (centre) and converted into oil-palm plantations (right).

Animal rescue Orangutans rescued from an 'amusement' park in Thailand are housed in temporary cages, awaiting a return home to better conditions – but still an uncertain future.

the Nigeria-Cameroon subspecies of chimpanzee extinct in the wild by about 2025: that is disregarding other factors such as habitat loss, human disturbance, and disease, which are likely to hasten the process appreciably. Logging has made many parts of the wild apes' forest home more accessible than ever before and that has allowed hunters to exploit new populations of prey. Abandoned logging camps are often used to provide accommodation for illegal hunters, and for a fee logging trucks will transport live or dead apes to the nearest commercial centre. Despite CITES bans on all traffic in great apes, there always seem to be people who will turn a blind eye if the price is right. Even in areas where wild apes themselves are not hunted for meat, they often fall victim to snares set for other animals such as antelopes. In one chimpanzee population in Uganda almost a quarter of the apes were found to have deformed limbs, almost certainly the result of their being caught in traps and snares. Other apes may be injured – even the powerful gorillas can easily lose a hand or a foot. Patrolling for and removing snares is a key feature of the work of conservation projects and has the added advantage of providing a human presence.

Disease

The spread of infection

One of the most dangerous effects of our close relationship with the wild great apes is how readily diseases can be transferred between us. There are more than 100 infections that can be passed from humans to wild apes or vice versa: diseases to which the gorilla, in particular, is vulnerable include the common cold, pneumonia, smallpox, measles, mumps, rubella, meningitis, and polio.

The potential for infection is a risk that, unfortunately, goes hand in hand with ecotourism. And debate continues to rage about the ethics of using non-human apes in research into our own diseases. Of all the diseases passed from apes to humans, the most devastating – and perhaps the least understood – is Ebola. Ebola haemorrhagic fever was first identified in 1976 in Zaire (now the Democratic Republic of Congo) and named after a river near where it was found. It is caused by the Ebola virus, and its natural reservoir (the place in which it originally occurs) is probably an animal that is native to the African continent, possibly a bat or rodent. The symptoms include severe headache, fever, diarrhoea, and, most seriously, internal and external bleeding. Death as a result of massive loss of blood occurs in 50–90 per cent of human cases and, although work on a vaccine is under way, there is as yet no cure. Over the past 30 years, outbreaks of Ebola among human populations in various parts of Equatorial Africa have killed perhaps 2000 people. The mortality rate in the wild great apes is believed to be even higher than in humans and Ebola has probably been responsible for the deaths of tens of thousands of gorillas and chimpanzees. In some places it has come close to wiping out whole populations, as happened with both gorillas and chimpanzees in parts of Gabon during two outbreaks in the 1990s.

Ebola has probably been responsible for the deaths of tens of thousands of gorillas and chimpanzees.

Ebola is spread by direct contact with an infected animal or human, or with their bodily fluids. An animal investigating the carcass of another, infected animal in the forest will be vulnerable, as will a mourner touching the body of a human victim during a burial ceremony. Handling or eating infected bushmeat is another obvious source of infection, which is not confined to contact with wild great apes – the virus can also be picked up from antelopes, porcupines, and other species. The only good news is that humans and apes that survive an Ebola attack seem to develop immunity.

The ecotourism dilemma

Habituating wild gorillas and orangutans to human company, so that tourists can enjoy a close encounter with them, is an important source of revenue for many countries. However, well-travelled visitors are likely to expose the vulnerable apes to a range of pathogens to which they have no immunity – even a cold or a cough can be fatal to a juvenile. A study of chimpanzees in the Taï forest of Côte d'Ivoire, published in 2008, showed that the presence of human researchers working with the wild great apes raised a significant risk of infection. On the other hand, the incidence of poaching was appreciably lower in areas where researchers were working. This poses a dilemma for conservation organizations: if we keep people away to stop the spread of disease, we increase the apes' chances of being trapped or shot. Balancing these concerns is one of the most significant challenges now facing conservationists.

While humans, being closely related to the wild great apes, carry the most pathogens that might infect them, ours is not the only species that poses a risk: diseases can be transferred from other animals too. An outbreak of anthrax, which killed at least eight chimpanzees in Côte d'Ivoire, for example, is thought to have been caused by contact with cattle being transported along routes through the forest.

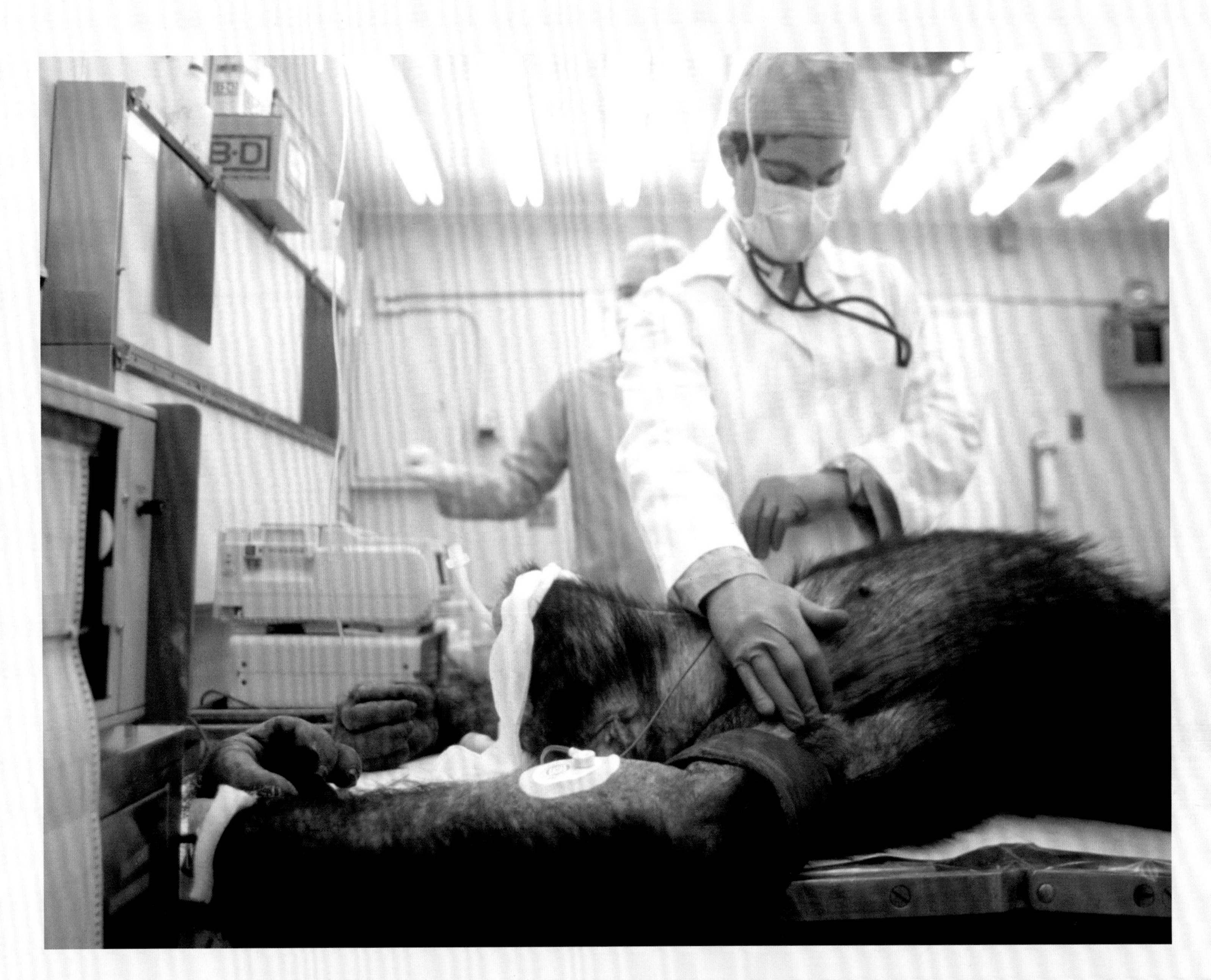

Hairier patient
Medical research using great apes has not yielded the hoped-for major breakthroughs for human health. It is now discontinued in the great majority of countries.

The research debate

Our close relationship with the great apes has also led to the widespread but controversial use of chimpanzees in research into human diseases, in recent years most notably AIDS. Scientists have known for some time that captive chimpanzees – and probably wild ones as well – carry an AIDS-related virus, Simian Immunodeficiency Virus or SIV, although this does not appear to develop into a disease as HIV does in humans.

Recent research seems to confirm that AIDS developed as a result of human contact with an SIV-carrying chimpanzee. Having established this fact has brought scientists a step closer to producing an AIDS-combating vaccine and, for some, this provides a powerful argument for the continuation of chimp-based research. That said, Great Britain, New Zealand, Sweden, and The Netherlands have all banned the use of great apes in research. Only the USA, Japan, Liberia, and Gabon still carry out research on chimpanzees. Today, the US National Institutes of Health support eight National Primate Research Centers, of which two breed and use chimpanzees. On the other side of the coin, The Humane Society of the United States runs a programme called Chimp Haven, which allows chimpanzees that are no longer used in research to 'retire' in comfort. That society continues to call for a total ban on using great apes in biomedical research.

Warfare

Human conflict affecting the great apes

In countries where large numbers of people have been driven from their homes or killed as a result of tribal conflict it is easy to overlook the impact that these human disasters might have had on wild great apes. But the fact is that human warfare does affect the wild animals that live in the places where it happens. And of the 23 nations that are home to at least one species of great ape, more than half have seen major conflicts in the last 20 years. All have had serious repercussions for their primate populations. . .

Human warfare impacts wild great apes in several ways. At the most basic level, when formerly productive farmland is turned into battlegrounds, food production falls away. Both soldiers and refugees still have to eat, and as a consequence the demand for bushmeat increases. At the same time, such conflicts increase the availability of guns, making it easier for people to kill wild great apes, as well as each other.

Without the 'tourist dollars' funds to pay for anti-poaching patrols and surveillance of animals in protected areas are simply not available.

From a financial point of view, conservation issues fall a long way down a government's list of priorities when money needs to be found for arms. Expatriate conservation organizations tend to move elsewhere when security issues become a cause for concern, and private investors also look for somewhere safer to spend their money. Tourists, too, stay away, and without the 'tourist dollars', the funds to pay for anti-poaching patrols and the surveillance of animals in protected areas are simply not available.

Disaster in DRC

In times of war, the less-than-scrupulous can always find a way of profiting from the situation. The Democratic Republic of Congo is rich in coltan, a black ore found in alluvial deposits along its rivers and mined by 'panning' in much the same way as people search for gold. Coltan has become valuable because it is used in miniaturized equipment for mobile phones, laptop computers, and other hi-tech electronic equipment. Unfortunately, the worldwide coltan boom, spurred by the popularity of these commodities, coincided with the devastating civil war of 1996–2002 in the Democratic Republic of Congo. In the six years that the war lasted, thousands of 'freelance' miners and their families flooded into the national parks, desperate to make what money they could. At the same time, an organized and well-armed band of professional hunters appeared, and proceeded to supply bushmeat of all kinds to feed the new mining community.

Even after the war had ended, wild great apes continued to be targeted by hunters in the Democratic Republic of Congo. During the first seven months of 2007, a total of eight adult mountain gorillas are known to have been killed for food in the Virunga National Park, one of them a female whose four-month-old baby was also found dead.

Under the circumstances that the country currently finds itself in it is difficult to provide accurate estimates of the total numbers of animals that have been lost. However, it is thought that the number of gorillas

Guarded apes For patrolling rangers in the Virunga region, poaching for bushmeat to feed local armed militias is just one of the multitude of threats facing the mountain gorillas.

Amid the fighting Despite continuing military action in the Virungas region, a late 2008 survey showed several babies born to the Mount Mikeno mountain gorilla groups. These gorillas seem to have been undisturbed enough to breed.

in the Democratic Republic of Congo may have shrunk from 17,000 in 1996 to as few as 4000 by 2002. The most optimistic assessment says that the bonobo population in the more accessible areas of its range declined by 25 per cent over the same period; figures for the impact of the conflicts on the local chimpanzee population are not available, but they too are likely to make gloomy reading.

Problems elsewhere

The war and subsequent events in the Democratic Republic of the Congo provide an extreme example of what has happened throughout western and central Africa in recent decades. In the early years of this century, civil war in Sierra Leone displaced more than a third of the human population and left many conservation areas unpatrolled, making it easy for hunters to shoot and trap wild chimpanzees. In Sudan, where bushmeat supplements the diet of those who have lost livestock during over 30 years of war, AK-47s abound. And during the civil war in the Democratic Republic of Congo in the late 1990s the situation became so extreme that Brazzaville Zoo was raided for meat to feed the starving populace. The zoo's great apes were only saved by being airlifted to safety by the workers for the John Aspinall Foundation and the Jane Goodall Institute.

Not enough genes

The problem with populations

A large population is important to the survival of any species, whether plant or animal, because it promotes genetic diversity. Shrinking numbers and the resultant loss of genetic diversity diminish a population's ability to adapt to change, which is in itself a threat to survival. Smaller populations are also at risk from unhealthy genetic mutations, which would normally be 'bred out' in a larger group.

Scientists define a species' minimum viable population as 'the number of individuals required for a high probability of survival of a population over a given period of time'. That definition is often arbitrarily made more specific and expressed as 'a better than 95 per cent chance of surviving for 100 years'. It is calculated by taking into consideration several factors, such as:

- inbreeding depression: that is, the extent to which inbreeding is likely to cause genetic problems in a given species;
- density dependence: the extent to which a population may suffer because of overcrowding or being too thinly distributed;
- catastrophes such as fire or flood, which may cause a sudden dramatic decline in population;
- length of time it takes an individual of the species to reach maturity. One of the problems with wild great apes is that they have a long 'childhood', with females reaching sexual maturity at the age of anything from six or seven (in some gorillas and orangutans) to twelve or more (in some bonobos). Another problem is the slow speed at which wild great apes reproduce. Generally, this is at four- to eight-year intervals, so if a population is struck by disease, for example, it cannot recover quickly.

A fragmented habitat

It is not easy to determine what a minimum viable population or MVP is for any given species without detailed study. However, just such a study has been carried out on the Sumatran orangutan. It reached the conclusion that a minimum viable population was 250–500 individuals, ranging over an area of 50–600 square km (20–230 square miles), depending on the quality of the habitat. In the lowland swamp forest of Suaq Balimbing in northern Sumatra, it has been extrapolated, an area of 100 square km (40 square miles) could support 725 orangutans, including 33 adult males, of whom only seven would be dominant – and therefore in a position to mate regularly and contribute significantly to the gene pool – at any one time. But the important point here is that there needs to be continuous, good-quality habitat in order for the orangutans to find enough food and to be able to range widely enough to avoid inbreeding. Habitat fragmentation is a potential threat to African apes too. The Cross River or Nigerian gorilla (a subspecies of the western gorilla) survives in tiny pockets in Nigeria and Cameroon, mainly in steep ridge forests above 400 metres (1,300 feet), where until recently inaccessibility protected it. Now human population pressure has left the apes with patches of forest sprinkled across land that has been cleared for farming and grazing. The world population is estimated at 250–280 weaned individuals, very much borderline for survival. In Côte d'Ivoire, a 2008 survey showed the chimpanzee population had collapsed by up to 90 per cent since the last survey 18 years earlier. At that time there were thought to be 8,000–12,000 chimpanzees in the country; now there are about 1,000, with only one viable population remaining. Major conservation efforts are required to reverse this trend.

One of the problems with great apes is they have a long childhood, so if a population is struck by disease it cannot recover quickly.

Cut off
Inhospitable agri-habitats may provide quick snacks for crop-raiding mountain gorillas. But these habitats also fragment the gorilla populations into unviable sizes where genetic diversity is insufficient for long-term survival.

Saving Planet Ape

A lot of the gloom and doom we hear about the future of the great apes is justified, and urgent action is needed to reverse current trends towards habitat loss. There are many organizations working to save our primate cousins, and many ways in which we as individuals can help. The future of the great apes in the wild may hang by a thread, but they can be saved if we can create sufficient international pressure to ensure certain regions of their natural homelands are protected. If we are complacent or indifferent, the wild apes will have vanished before the end of this century.

Global efforts

What international organizations are doing

The list of global organizations dedicated to the conservation of wild creatures and the environment is an impressive one: IUCN, CITES, UNEP and its great-ape-specific-programme (GRASP), WWF, Conservation International, the African Wildlife Foundation, Fauna and Flora International – the list goes on and on, and many millions of pounds, dollars and euros are poured each year into projects around the world. But who are these organizations, and what do they really achieve?

International initiatives

The International Union for Conservation of Nature (IUCN) supports research and manages field projects all over the world; it has more than a thousand government and NGO member organizations, and almost 11,000 volunteer scientists in more than 160 countries. Most famously, it produces the Red List of Threatened Species, a guide to determining which wild creatures are most in need of conservation action.

Many of the protected areas where great apes live have been designated World Heritage Sites by UNESCO.

The Convention on International Trade in Endangered Species of Wild Fauna and Flora (CITES) is an international agreement between governments whose aim is 'to ensure that international trade in specimens of wild animals and plants does not threaten their survival'. CITES gives varying degrees of protection to more than 30,000 species of animals and plants, whether they are traded as live specimens, fur coats, or dried herbs. All the great apes are listed in Appendix 1 of the Convention, which means that international trade in them is permitted only in exceptional circumstances, such as the exchange of animals between zoos as part of a reputable captive-breeding programme (see pages 280–81).

Many of the protected areas where great apes live have been designated World Heritage Sites by the United Nations Educational, Scientific and Cultural Organization (UNESCO). As the name suggests, World Heritage seeks to encourage the identification, protection and preservation of areas 'of cultural and natural heritage around the world considered to be of outstanding value to humanity'. The proposal to create World Heritage Species – to heighten awareness of animals that are also of 'outstanding value to humanity', which would include all the great apes – has not yet come to fruition.

The Forest Law Enforcement and Governance (FLEG) process is a global effort to address the violation of laws protecting forests, with the aim of balancing short-term remedies and long-term, sustainable solutions involving local communities. Unfortunately, the words 'requires massive monitoring, training and law enforcement' recur frequently in its list of proposals. It is a sad fact of life that logging companies are often better armed and better equipped than park rangers, and that in the 35 Indonesian national parks for which figures are available 2,155 rangers are expected to patrol an area of 108,000 square km (over 40,000 square miles). That is over 50 square km (20 square miles) each, assuming they worked individually, which of course would make them very vulnerable to attack by poachers or others they were trying to police.

The Great Apes Survival Programme was founded by UNEP (the United Nations Environment Programme) and UNESCO with a mission to work at the highest level to promote cohesion of policy and to maximize efficiency, communication and effective use of resources between governments, NGOs, conservation organizations and others working for the good of the great apes. However, the first inter-governmental meeting of GRASP, which established the 'political will' to secure the future of the great apes till 2015, did not take place until 2005, four and a half

◄◄ Surrogate mothers (Previous page). Ape aid workers in East Kalimantan, Indonesia, take detailed care of their orphaned orangutan charges, who lack the formative years of one-to-one maternal teaching and guidance.

◄ Orangutan orphanage
Borneo Orangutan Survival is dedicated to saving orangutans from extinction and protecting their rainforest homes. BOS's main focus is the rescue and rehabilitation of hundreds of wild and orphaned Bornean orangutans.

years after the project's official founding. During which period, according to one expert, perhaps 20,000 orangutans died. Shortly before this conference, the minister in charge of tourism and the environment in the Democratic Republic of Congo authorized a system of licensed hunting whereby a mountain gorilla could be shot for a fee of $500 and a live one captured for $1000. The payment of 'an appropriate fee' also permitted the payer to catch or shoot bonobos and chimpanzees. This is despite the fact the GRASP meeting took place in the Democratic Republic of Congo, and that the country is a (voluntary) party to the CITES agreements.

It's a fact of life that logging companies are often better armed and better equipped than park rangers.

Other international initiatives have also been slow to get off the ground. In the late 1980s and early 1990s, the European Commission took five years to put in place the finance to support a project to develop the Cross River National Park in Nigeria. By the time they had awarded the management contract to the firm of their choice, the Nigerian government – despite having huge political and economic difficulties to contend with – had gone ahead and created the park. A lot of 'hands on' conservationists feel there is plenty of assessment of the situation going on, plenty of lip service being paid to the ideals, but not enough action to deal with what everyone recognizes is a crisis requiring urgent solutions – and large-scale, reliable, long-term funding. A real breakthrough for the bonobos came in 2007 when the Democratic Republic of Congo announced the creation of the Sankuru Nature Reserve, protecting some 30,000 square km (almost (12,000 square miles) of rainforest. But, as the president of the Bonobo Conservation Initiative said, 'Our work has just begun. Now we need investment to successfully manage the reserve.'

Charity work

Some conservation success stories

Despite setbacks, conservation work has made great progress. Thirteen protected areas containing chimpanzees, ranging across 11 countries, were surveyed in a study published in 2004 which identified the factors contributing to conservation success. In order of importance, the top four were: a positive public attitude, effective law enforcement, large protected area size, and low human population. Add to that a change in attitudes in the West – to reduce consumption in general, and in particular demand for illegal timber, both of which contribute so much to climate change – and you have a summary of the areas on which most wildlife conservation organizations concentrate.

The panda logo of WWF – the Global Conservation Organization (still called World Wildlife Fund in the US) – is by far the most recognizable symbol of wildlife conservation worldwide. Launched in 1961, WWF now operates in over 90 countries and works with governments, businesses, and individuals to 'address global threats to people and nature such as climate change, the peril to endangered species and habitats, and the unsustainable consumption of the world's natural resources'.

The observations compiled by scientists have vastly improved our understanding of ape behaviour and social structure.

Conservation International (CI)'s mission is 'to conserve the Earth's living heritage – our global biodiversity – and to demonstrate that human societies are able to live harmoniously with nature'. Since its launch in 1987 it has worked with multinationals such as Starbucks and Wal-Mart to promote conservation issues, and with governments across the globe 'to make conservation a core component of their national policies'.

In 2007, the WWF was party to a potentially historic agreement between the governments of Indonesia, Malaysia, and Brunei Darussalam to protect, manage, and restore a 220,000 square km (85,000 square miles) stretch of forest in the Heart of Borneo – that's an area almost as large as Great Britain, about the size of the state of Utah. WWF has forest projects in East Africa, too, and it is one of the co-funders of the International Gorilla Conservation Programme, which somehow continued to support the protection of gorillas in Rwanda during the genocide period of the 1990s. Fauna and Flora International (FFI) and the African Wildlife Foundation (AWF) are also involved in this programme. CI, again in partnership with FFI, implemented the Liberia Forest Reassessment programme and has been working on an education and awareness project to improve the lot of the chimpanzee in West Africa. CI is also responsible for identifying the 34 'biodiversity hotspots' mentioned on pages 250–1 as being in particular need of conservation attention, while AWF helped to create the protected area of the Lomako Forest in the heart of bonobo country. Among FFI's other projects are the training of patrol units to protect orangutan habitat in Borneo.

One of the most effective roles these international organizations can play is in liaising with governments to establish and maintain 'transfrontier protected areas' to cover the many areas where great ape territories cross national borders. The EU is involved in such a project

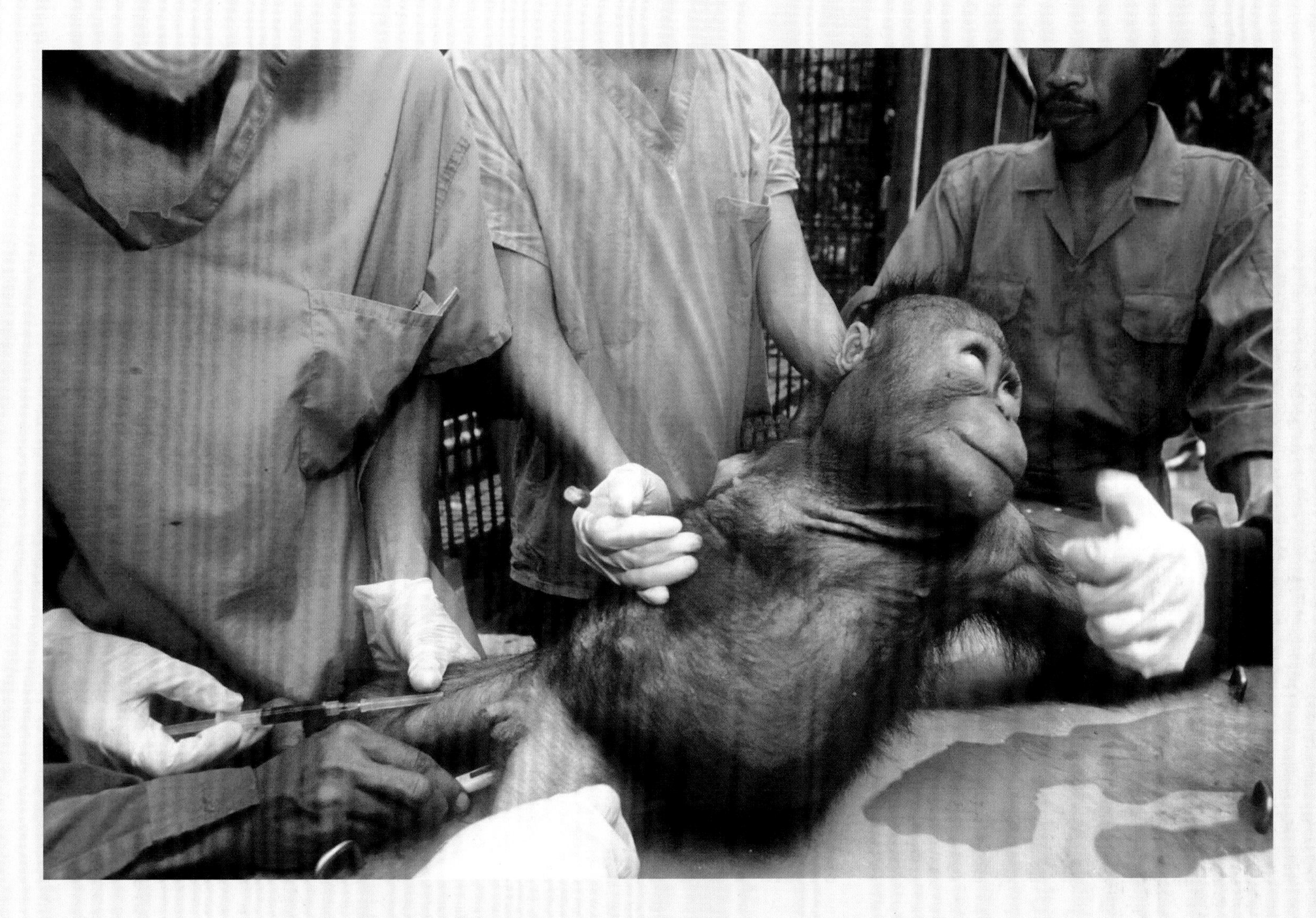

Safe on an island Chimpanzees at the Ngamba Island Sanctuary in Lake Victoria receive veterinary care. The 40-hectare (100 acre) sanctuary cares for orphaned chimps who would be unable to survive if reintroduced into the 'proper' wild.

Ape-to-ape care A young Bornean orangutan receives a monthly medical check-up. Workers are always on guard against human-introduced or natural orangutan infections that might spread through the sanctuary.

to protect a continuous area of chimpanzee habitat in West Africa, while WWF-Indonesia is working to the same end in Borneo, which part belongs to Indonesia, part to Malaysia.

The reasons for research

Since Jane Goodall started watching chimpanzees at Gombe in 1960, long-term studies of the great apes have spread all over their range. The observations compiled by scientists have vastly improved our understanding of ape behaviour and social structure. The projects have attracted publicity and funding, raised international awareness of conservation needs, provided local employment, and – by virtue of the mere presence of researchers – helped to deter poachers and illegal operators of all kinds.

Additional knowledge of animal behaviour can be passed on to tourist guides and enrich the visitors' experience; it can also persuade the authorities that a protected area should continue to be protected, or that an unprotected area deserves to be. But perhaps the field researchers' most important achievement is to increase our understanding of habitat requirements. We know much more than we did half a century ago of the importance of continuous stretches of pristine forest to apes that range over an area of 25 square km (nearly 10 square miles). One breakthrough in observations of western gorillas came as late as the 1990s when scientists discovered baïs, clearings around watercourses where gorillas come to eat, drink, bathe, and play. Being able to watch the gorillas behaving naturally in lighter and more open habitat than their usual dense forest home gave new insights into gorilla social life, group formation, and interaction between groups, all of which impacts on their conservation needs.

Ecotourism and community efforts

Conservation initiatives on a smaller scale

Protecting a wildlife area is not only an expensive business, it simply cannot be done without the goodwill of people 'on the ground'. While wealthy Westerners may be prepared to pay $500 for the privilege of spending an hour with a group of mountain gorillas, the local people must see some benefit if this form of tourism is to succeed in the long term.

People who are asked not to graze their livestock or gather firewood from an area that is to be given over to wildlife must be compensated in some way if this is not to cause friction. Similarly, they cannot be expected to tolerate wildlife damaging their crops if they gain no advantage from the presence of that wildlife. As the biologist Jonathan Kingdon wrote in his book *Island Africa*, '...the primary need is for the opening of friendly dialogues, partnerships and a sharing of knowledge and enthusiasm with the people, especially the poorer inhabitants... Conservation here is primarily a social, political and human problem.'

The most sustainable conservation projects, therefore, work outwards or 'from the bottom up,' starting with the local communities. One immensely encouraging example of this occurred in 2004 in northern Sumatra, where the Batang Gadis National Park was created in response to pressure from the community, who were protesting against the environmental damage caused by an increase in illegal or corruptly licensed logging. This means that an area of over 1,000 square km (400 square miles), which contains one of the world's most diverse flora and has historically been home to the Sumatran tiger and rhinoceros as well as the orangutan, is destined to become part of a huge northern Sumatra biodiversity conservation corridor – because 30,000 local people wanted it to happen!

An alternative to bushmeat

It has been said that people consume bushmeat because they need protein and sell it because they need money. Although the bushmeat trade is now known to operate on a much larger scale than this simple assessment suggests (see page 262), community-based projects and integrated conservation and development projects (ICDPs) aim to address these issues by providing both income – much of it from tourism, as described below – and alternative sources of protein. Sustainable farming for cash as well as subsistence, and beekeeping to meet an increasing worldwide demand for honey are two options; others include farming alternative sources of protein in the form of fast-breeding animals such as chickens and rabbits, or non-endangered wild species such as the cane rat and the red river hog. In coastal regions, fish farming is also promoted. Public awareness campaigns, promoting the value of the great apes alive rather than dead, have brought about a marked reduction in the consumption of bushmeat in some parts of West Africa. The brief of ICDPs extends to such issues as environmental education, clean water supply, health care, and family planning. The hope is heightened awareness of conservation issues, grass-roots disapproval of illegal hunting, alternative sources of food and income, and more efficient enforcement of the laws protecting wildlife and its habitats will lead to a reduction in hunting and habitat pressure on great apes.

The most sustainable conservation projects work outwards or 'from the bottom up' starting with the local communities.

Ecotourism

Ecotourism is undeniably lucrative: a non-Rwandan resident pays $500 for a day pass to the Volcanoes National Park; that is about one and a half times the country's average annual per capita income. Tracking gorillas in Uganda is equally expensive, while a four-night 'chimpanzee safari' at Gombe Stream in Tanzania costs more than $2,000. Increasingly – in theory, at least – much of this money is poured into further conservation projects.

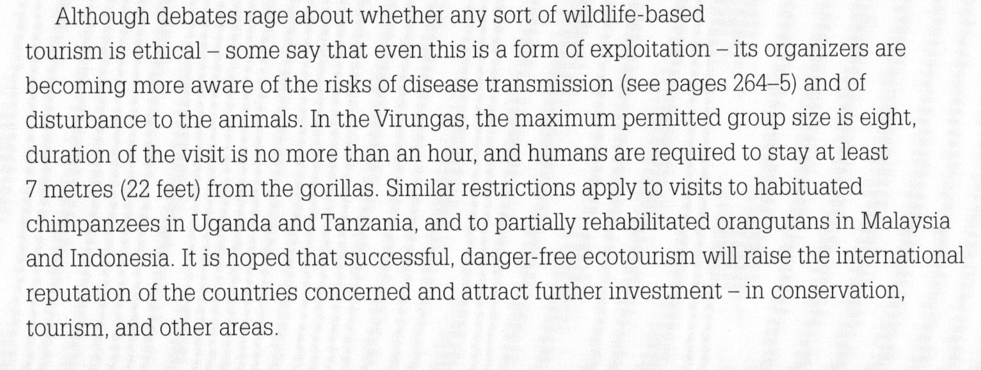

Tourism also provides jobs: in hotels, restaurants, and safari camps, as park rangers and guides, as makers and sellers of souvenirs. One of the concerns here, however, is the danger of relying on tourist income in areas that are prone to outbreaks of violence or disease: few visitors went to Rwanda during the genocide of the 1990s; few are venturing into the Democratic Republic of Congo at the time of writing. Unrelated external factors, too, can have a negative impact: it is safe to assume that the 2008 'credit crunch' in the developed world will considerably reduce the disposable income of many eco-travellers.

Ecotourism is undeniably lucrative: a non-Rwandan resident has to pay $500 for a day pass to the Volcanoes National Park.

Although debates rage about whether any sort of wildlife-based tourism is ethical – some say that even this is a form of exploitation – its organizers are becoming more aware of the risks of disease transmission (see pages 264–5) and of disturbance to the animals. In the Virungas, the maximum permitted group size is eight, duration of the visit is no more than an hour, and humans are required to stay at least 7 metres (22 feet) from the gorillas. Similar restrictions apply to visits to habituated chimpanzees in Uganda and Tanzania, and to partially rehabilitated orangutans in Malaysia and Indonesia. It is hoped that successful, danger-free ecotourism will raise the international reputation of the countries concerned and attract further investment – in conservation, tourism, and other areas.

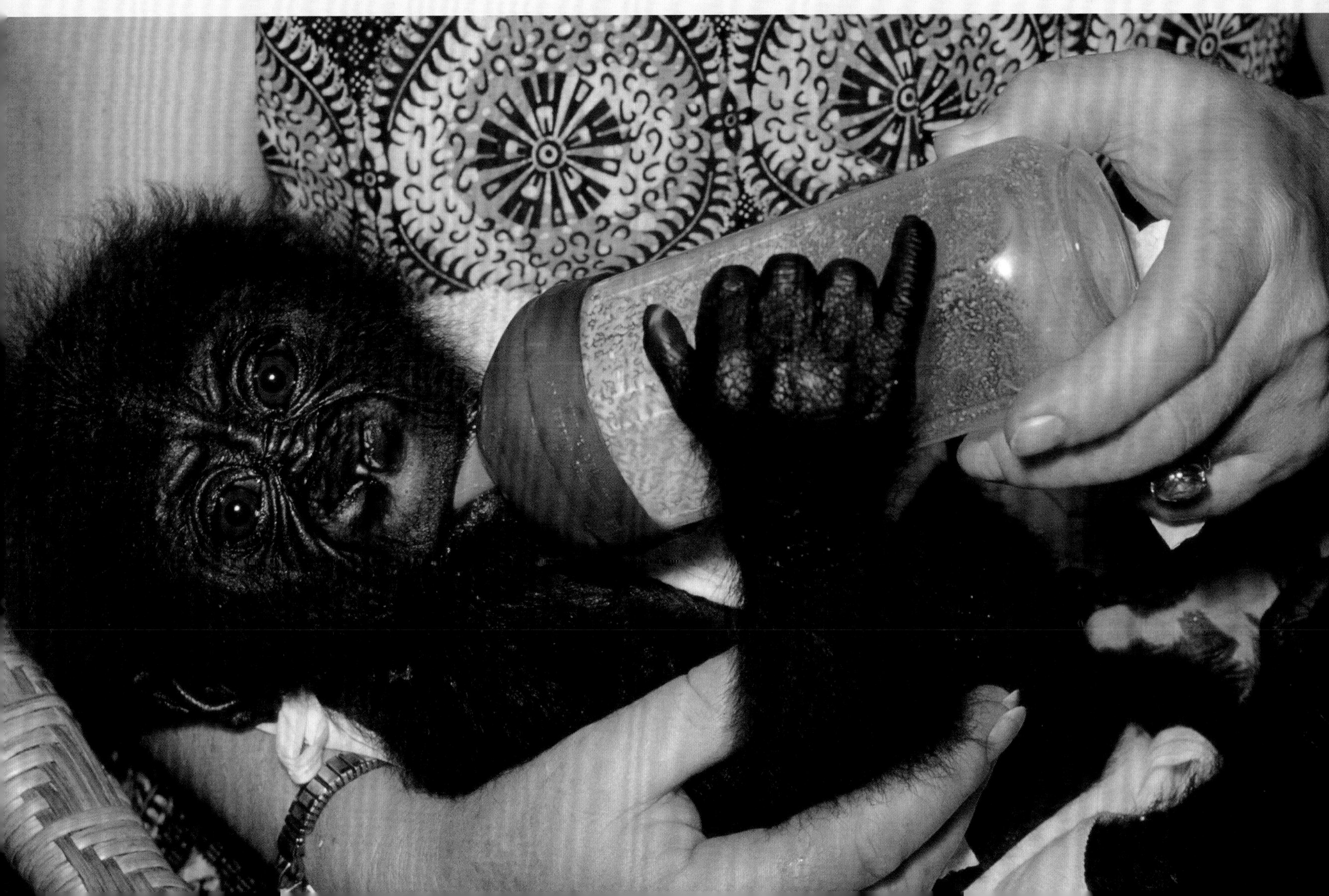

On the bottle
Orphaned great apes like this baby gorilla can be fed modified human formula milk in an effort to regain their condition and strength. However the potential for reintroduction into the wild depends foremost on the attitudes of local people.

Captive-breeding

An important role for zoos

For many years arguments have raged about the ethics of zoos and wildlife parks. Their detractors believe the whole concept of keeping animals for exhibition is exploitative; and there will always be bad zoos which keep animals in degrading conditions. But the better ones – of which there are now many across the world, from London to San Diego, Frankfurt to Melbourne – have important captive-breeding programmes which help to maintain genetic diversity in endangered species.

Well-designed zoos enable the animals to express natural behaviour in good conditions, so that visitors can see them enjoying themselves or at least getting on with their lives, rather than pacing up and down in the cramped cages that were the norm a generation or so ago. The educational and 'ambassadorial' role of enabling people from all over the world to become acquainted with wild creatures should not be overlooked. As part of their captive-breeding programmes, zoos across Europe, America and Australasia regularly 'swap' animals, carefully monitored to avoid inbreeding. (This is one of the 'exceptional circumstances' in which CITES permits the international transportation of great apes.) Animals may be microchipped, fingerprinted, and provided with passports as a guard against illegal trading. Captive populations also enable further research into the apes' social habits, psychology, and causes of illness and death.

Reintroduction

Many zoos liaise with what are called in situ conservationists, providing money, professional expertise, and sometimes animals for release back into the wild. The Frankfurt Zoological Society, for example, also works closely with the Congolese Wildlife Authority to protect the gorillas of Virunga National Park; the gorilla even forms part of the zoo's logo. In Australia, Zoos Victoria – of which Melbourne Zoo is a part – works with Fauna and Flora International to protect part of the Gunung Leuser National Park in northern Sumatra, an important orangutan habitat. Reintroduction to the wild has not yet proved very successful with captive-bred great apes, though early signs are promising for the John Aspinall Foundation's western gorilla programme in Gabon and D. R. Congo. In 2007, seven years after the first introductions, the survival rate of the captive-born apes was 86 per cent (remarkably high), and the group's next generation baby had been born. But given the gorilla's slow reproduction rate, it is still too soon to make long-term predictions. See also pages 282–3.

Well-designed zoos enable animals to express natural behaviour in good conditions, so that visitors can see them getting on with their lives.

Translocation

An alternative to reintroduction, translocation involves taking members of small, isolated populations of a species and moving them to a new location where they can socialize and interbreed, thus increasing their gene pool and producing more viable communities. The other benefit – its supporters argue – is that a single population gathered together in a single, albeit larger, location is easier to protect. Translocation has been suggested for the Cross River gorillas (see page 268), whose fragmented habitat makes them textbook candidates for this approach. However, the cost of veterinary care while the animals are in transit and the high risk of disease being transmitted either from one ape population to another or from apes to humans and vice versa mean that this is unlikely ever to become a 'mainstream' conservation strategy.

Halfway house (Previous page) Young orangutans, between about two-and-a-half and five years of age, are limited to a 10-hectare (20 acre) walled area or 'high school' where they learn the ways of the wild before full-scale reintroduction.

Unsure outlook Breeding programmes of great apes, such as these western lowland gorillas at Zoo Atlanta, USA, are carefully coordinated with successful matings and offspring recorded on global databases.

Rehabilitation and reintroduction

A new life for rescue orphans

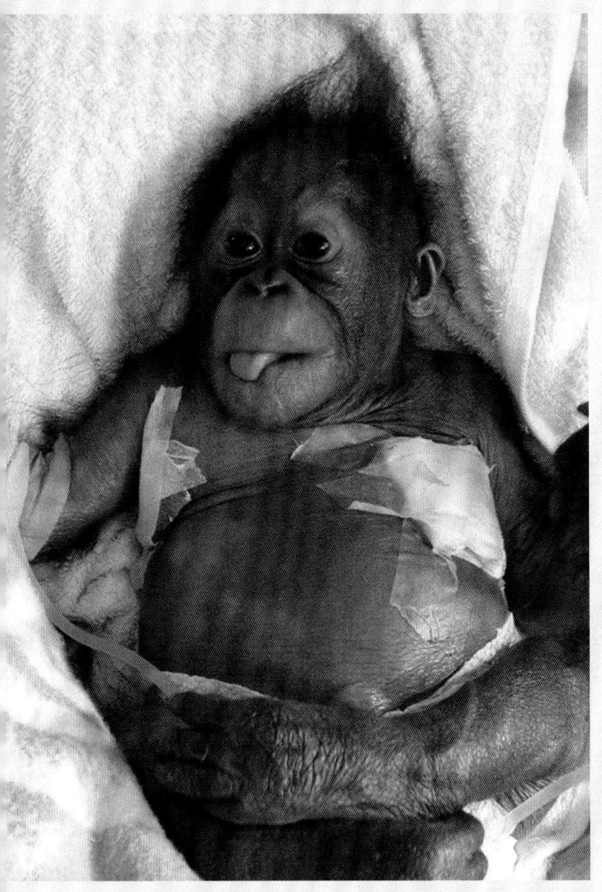

Reintroducing great apes to the wild has generally been more successful with wild-born orphans than with captive-bred animals. In the 1970s, an attempt by Stella Brewer Marsden, founder of the Chimpanzee Rehabilitation Trust, to reintroduce orphaned chimps in Senegal failed: pushed into close proximity by a severe drought, the wild chimpanzees attacked the incomers. Later attempts were more successful, and the Trust now monitors a population of 76 chimps that it has introduced to three islands in the River Gambia National Park – three infants born recently are the grandchildren of the original orphans confiscated from hunters and traders and released in the 1980s.

Releasing chimpanzees into mainland forest areas is more problematic, but has been achieved in the Conkuati-Douli National Park in the Democratic Republic of Congo. This project, known as HELP Congo, concentrates on releasing females, which it believes are more likely to be accepted by wild males and go on to reproduce successfully.

At Petit Evengué in Gabon, orphaned gorillas are rehabituated to the forest. The youngsters spend time in the forest every day with their keeper, getting to know how to look after themselves; they will be released at soon as a suitable habitat is secured.

The question of habitat is key to successful rehabilitation and release programmes. Great care is needed to ensure that the apes do not harbour any human infections, and that they are not released anywhere where they will be treated as intruders by resident apes. Nor can they be introduced too close to human settlement: animals that are used to treating humans as a source of food and comfort are likely to wander into gardens, destroy crops, and become aggressive if they are not fed. Official policy in Indonesia is to release orangutans into areas where populations are known to have been lost, but even this requires careful monitoring of logging activities, hunting, and other potential pressures: there is no point in releasing animals back into the conditions that led to them being orphaned in the first place.

There is no point in releasing animals back into the conditions that led to them being orphaned in the first place.

All of which again raises the vexed question of finance. The Borneo Orangutan Survival Foundation (BOS), for example, cares for up to a thousand orphaned orangutans at a time and has released over 450 into the wild. BOS's 'residents' are mainly youngsters captured for the pet trade when their mothers are shot for meat; by the time they reach the sanctuary they are usually injured and starving. BOS does its best to nurse them back to health and teach them the skills their mother would have taught them, from climbing trees to making friends. None of this comes cheap: BOS has been known to spend $10,000 caring for a single, badly injured orphan.

Many programmes continue to feed and/or monitor the apes after release: the HELP project will monitor theirs until 2013. This obviously provides much-needed data on the success of the reintroductions (not least to enable people to judge whether it is worth continuing to fund such projects), but also involves further expense.

Keeping calm
Sedation for travel or medical checks can result in temporary 'anaesthetic hangover' that affects muscle coordination – including the tongue.

Wild school
Orangutans and their carers share excitement as they prepare for a day's climbing practice, as part of their path to reintroduction.

Funding

Critics of ape sanctuaries and release programmes cite any or all of the above difficulties, but as NGOs working with great apes have the legal authority (not to mention the moral obligation) to confiscate animals that are being smuggled or illegally captured, housing them somewhere is the only alternative to euthanasia. Enforcing laws more tightly in countries such as Taiwan, where young apes have traditionally been popular as pets, has led to more and more youngsters being rescued from illegal captivity and needing care. Donations and legacies to charities and NGOs are essential if this work is to continue.

Another way of providing funding is a 'debt for nature' scheme which is being given a new slant in the proposed Mawas National Park in Indonesia, one of the most important orangutan habitats left in the world. This 500,000 hectare (1,930 square miles) forest is currently a protected reserve, and the future of the project relies on the area being upgraded.

The idea behind debt for nature is simple: a country agrees to protect a wild area in perpetuity, in return for being 'forgiven' some of its foreign debts. However, in order for the scheme to work, the country in question has to honour its pledge to protect the area – and to do it effectively.

How we can help

Our contributions make a difference

The rich West's consumption of products from the rainforests drives much of the habitat destruction described in the previous chapter. Without a fundamental change in our attitudes and buying habitats, the apes' homes and the hugely rich diversity of animals and plants that the forests support face continued decline and probable extinction. At the current rate of destruction, all the well-drained lowland forest of Kalimantan in Borneo is likely to be lost by 2018 at the latest – and with it will go about half of the Bornean orangutan's range.

If we in the West simply stopped buying wood and wood products of unknown provenance, the market for illegal logging would dry up. The Forestry Stewardship Council (FSC) now certifies over 100 million hectares (400,000 square miles) of forest worldwide. That is 7 per cent of the world's productive forests, and the figure is increasing all the time – national initiatives in Nepal, Norway, and Honduras were accredited in December 2008. The FSC label assures the consumer that products come from forests 'that are managed to meet the social, economic and ecological needs of present and future generations'; FSC also helps businesses to reach ecologically aware markets.

So, the first thing that we as consumers can do is not to buy tropical wood products that do not bear the FSC label – and tell retailers or distributors why.

As eco-tourists, we have a responsibility to obey the rules and to treat the animals with respect.

We can vote with our feet in other ways, too. We can refuse to visit zoos that do not keep animals in as nearly as possible natural conditions; stay away from circuses or other entertainments that use apes; boycott airlines involved in the illegal wildlife trade or products that demean apes by using them in their advertising. And we can generally try to consume less: forests are being destroyed to produce coltan for use in mobile phones, so upgrading our phones less frequently is a small but important gesture.

As consumers we should also be aware of the way that palm-oil products have crept into our supermarkets over the last decades. The aims of the Roundtable on Sustainable Palm Oil (RSPO) include the promotion of the certification of sustainable palm oil in a similar scheme to FSC certification. This movement is gaining momentum: all the major UK supermarkets and multi-nationals such as Unilever are signatories, and Malaysia has allocated 50 million ringgit (about $13.5 million) of its 2009 budget to help smallholders gain RSPO certification. Organizations such as BOS are encouraging members of the public to be more proactive in demanding that products containing palm oil be sustainable. As with wood, we need to shop selectively and make retailers aware of why we are making our decisions.

Ecotourism

Visiting wild apes in reputable zoos, parks, and sanctuaries around the world can be of benefit. Our entrance fees and the money we spend on souvenirs help support conservation projects. As ecotourists, though, we have a responsibility to obey the rules and to treat the animals with respect – to remember that these are wild creatures, not babies to be cuddled, and that they are vulnerable to disturbance, infection, and stress. We should always do as a guide or ranger tells us, leave when our allocated time is up, and not sneak off the prescribed route to get just one last photo.

Charities and NGOs

The conservation organizations mentioned throughout this chapter need constant and – importantly – predictable supplies of money if they are to continue with their work. Most are delighted to receive even a small monthly sum (by direct debit or standing order) as it enables them to budget and plan expenditure. One of the most rewarding ways of making this contribution, particularly for children, is by adopting an animal. Supporters then receive regular updates on the animal's progress and can feel that their contribution is genuinely making a difference. WWF (wwf.org.uk/adoption/), the Jane Goodall Institute (www.janegoodall.org/chimp_guardian), and BOS (www.savetheorangutan.co.uk) are among the many who run this sort of scheme. Joining Give As You Earn, which allows a monthly donation to be deducted at source from an employee's salary, is easy and tax-efficient. One-off donations and remembering an organization in a will are also possibilities.

If we join any of these organizations (and details of how to help are on all the websites), we will be encouraged to write to our MP, write to major retailers, join campaigns to heighten public awareness, become involved in fundraising, sign and encourage others to sign petitions… The BOS website also mentions 'inappropriate' film clips on YouTube showing orangutans being used for entertainment and asks users to post complaints. If we want to help, there are many ways to do it, and many of them ask us to give only a few minutes of our time without leaving our living room.

Unusual patients The 'use' of chimpanzees and other great apes for medical and biological experimentation has greatly reduced in recent years. However the knowledge gathered there helps vets/doctors – to treat captive apes when they fall sick.

Charities & organizations

African Wildlife Foundation

www.awf.org

Operating in the African heartlands from Uganda and the Democratic Republic of Congo to South Africa, managing the vast swathes of wild land that is home to Africa's wildlife.

Ape Alliance

www.4apes.com

An international coalition of organizations, and individuals, working for the conservation and welfare of apes. Providing a forum for the discussion of issues relating to apes, and an effective body for lobbying and campaigning.

Bonobo Initiative

www.bonobo.org

Dedicated to ensuring the survival of the bonobo and its tropical forest habitat in the Congo Basin.

Born Free Foundation

www.bornfree.org.uk

Take action to protect threatened species and stop individual animals suffering. Work with local communities to help people and wildlife live together without conflict. High-profile campaigns help change public attitudes and persuade decision-makers to respond quickly to emergencies.

Bornean Orangutan Survival

www.savetheorangutan.org

Borneo Orangutan Survival (BOS) Foundation operates the largest primate rescue centre in the world, and provides opportunities for people to take part in shared adoption programmes.

Cameroon Wildlife Aid Fund

www.cwaf.org

CWAF runs the Mvog-Betsi zoo and Mefou National Park in Cameroon, in conjunction with the Cameroon government. Work with the government, local communities, and ecological groups around the world to protect Cameroon's wildlife.

Center for Great Apes

www.centerforgreatapes.org

Providing a permanent sanctuary in Florida for orangutans and chimpanzees who have been retired from the entertainment industry, from research, or who are no longer wanted as pets.

Chimpanzee Rehabilitation Trust

www.chimprehab.com

Set up in 1969 by Stella Marsden, as a solution for a group of chimpanzees confiscated from hunters and traders by the Gambian wildlife authorities, it is the longest running chimpanzee rehabilitation project in Africa.

CITES (Convention on International Trade in Endangered Species of Wild Flora and Fauna)

www.cites.org

An international agreement between governments. Its aim is to ensure that international trade in specimens of wild animals and plants does not threaten their survival.

Conservation International

www.conservation.org

Working to save species, conserve landscapes and seascapes, empower local communities, raise awareness, and develop innovative methods of conservation worldwide.

Dian Fossey Gorilla Fund International

www.gorillafund.org

Founded by Dian Fossey in 1978 and dedicated to the conservation of gorillas and their habitats in Africa through anti-poaching, monitoring, research, education, and support of local communities.

Fauna and Flora International

www.fauna-flora.org

Founded in 1903, it was the world's first international conservation organization and the pioneering work of its founders in Africa led to the creation of numerous protected areas, including Kruger and Serengeti National Parks.

Friends of Bonobos

www.friendsofbonobos.org

US charity that supports Lola ya Bonobo, the sanctuary of the NGO, Les Amis des Bonobos.

Gearing up 4 Gorillas

www.g4g.co.uk

Aim of 'Gearing Up 4 Gorillas' (G4G) is to help secure the long term survival of the critically endangered mountain gorilla, through providing practical assistance to rangers on the ground.

Great Ape Action Fund (IUCN/SSC Primate Specialist Group)

www.primate-sg.org/sga

Active throughout the tropical world, working in dozens of nations in Africa, Asia, and Latin America, the PSG promotes research on the ecology and conservation of hundreds of primate species — monkeys, apes, lemurs, and their many relatives.

Great Apes Survival Partnership – GRASP (UNEP/UNESCO)

www.unep.org/grasp/

GRASP's role is to complement existing great ape conservation efforts, through intergovernmental dialogue and policy making, conservation planning initiatives, technical and scientific support to great ape range state governments, flagship field projects and fund and awareness raising in donor countries.

IFAW (International Fund for Animal Welfare)

www.ifaw.org

Aim to improve animal welfare, prevent animal cruelty and abuse, protect wildlife, and provide animal rescue around the world. Projects range from stopping the elephant ivory trade, to ending the Canadian seal hunt and saving the whales from extinction.

H.E.L.P. Congo

www.help-primates.org

Habitat Ecologique et Liberte des Primates provide homes for orphaned chimpanzees, acting to protect primates, their habitats and their future existence. Also provide a chimpanzee adoption scheme.

IUCN (International Union for Conservation for Nature)

www.iucn.org

Conserve biodiversity through aiming to monitor climate change, energy, protect livelihoods and impact on economic factors affecting the environment.

Jane Goodall Institute

www.janegoodall.org.uk

Working primarily with wild chimpanzees, Jane Goodall is one of the world's most famous scientists, whose research in Africa continues to this day and is the longest field study ever undertaken of any group of animals in the wild. Today the institute combines scientific work in the field with international advocacy on behalf of chimpanzees and the environment.

Orangutan Appeal UK

www.orangutan-appeal.org.uk

A registered charity based in the south of England, dedicated to the rehabilitation and preservation of orangutans and the conservation of their habitat.

Strive to protect remaining wild populations of orangutans by providing support and funding for projects across Malaysian and Indonesian Borneo; and by raising awareness of their plight across the globe. The Appeal is also authorised to work on behalf of the famous Sepilok Orangutan Rehabilitation Centre!

Orangutan Foundation

www.orangutan.org.uk

Working across the entire range of the orangutan species to conserve the orangutan and the biodiversity of their habitat through the protection of the tropical forests of Borneo and Sumatra.

WSPA

www.wspa.org.uk

Focused on animal welfare areas including cruelty prevention and prevention of commercial exploitation of wildlife and killing of wild animals for food or by-products. Also disaster management – providing care to animals following man-made or natural disasters, and thereby protecting people's livelihoods.

WWF

www.wwf.org.uk

World's leading environmental organization founded in 1961 and active in over 100 countries. With their unique combination of practical experience, knowledge and credibility, work with governments, businesses and communities around the world so that people and nature thrive within their share of planet's natural resources.

Acknowledgements

Mitchell Beazley would like to acknowledge and thank the following for photos, with special thanks to BOS and Dian Fossey for their contribution.
(c-centre, b-bottom, t-top, r-right, l-left)

Alamy Alan Curtis 142 cr, Arco Images GmbH 233, 240 a, blickwinkel 254, Jacques Jangoux 143 cl, Mary Evans Picture Library 68 al, Moonbrush 142 cl, Papilio/Robert Pickett 234-5, Suzy Bennett 244 **Ardea** Adrian Warren 119 l, 138 br, 139 bl & r, 162-3, Andrey Zvoznikov 55 b, Duncan Usher 196, Ferrero-Labat 173 bl, 205, Ingrid van den Berg 54 b, Jean Michel Labat 22, 24-5, 238, Jean Paul Ferrero 170-1, 204, John Cancalosi 120 r, 197 b, 222-3, Kenneth W Fink 31, 182, M Watson 17 ac, 54 a, 105 b, 128, 157, 245, Masahiro Iijima 62, Nick Gordon 52 b, 53 b, Robin Stewart 65, Yann Arthus-Bertrand 18 br, 39, 114 br, 142 br
BOS F Borneo Orangutan Sam Gracey 174, Simon Bell 78-9
Corbis Andy Rouse 120 l,144, DLILLC 51 cb, Eric & David Hosking 121, Frans Lanting 5, 14, 51 ca, 51 a & b, 58-9, 61, Jan Butchofsky-Houser 53 a, Karl Ammann 87 al & ar, Keren Su 75, Kevin Schafer 52 ca, 55 c, Martin Harvey 52 cb & a, 57 a, Stan Osolinski 101, 135, 239, Theo Allofs 53 ca, Wolfgang Kaehler 50 b, Andy Rouse 124 l, Bettmann 69 ar, 83 cr, 85, 93 al, ac, & ar, 94, Fancy/Veer 240 b, Gallo Images 102, 186, George D Lepp 118 br, Hulton-Deutsch Collection 77, 83 ar, Jan Butchofsky-Houser 16 cb, Jessie Cohen/Smithsonian's National Zoo/Reuters 252, Larry Williams 241 bl, Penny Tweedie 274, Robert Maass 265, 285, Sukree Sukplang/Reuters 263 **FLPA** Cyril Ruoso/Minden Pictures 16 ca, 18 al, 43, 44, 48, 55a, 115 bl, 138 al & ar, 139 al, 143 bl, 173 a, 175, 192, 203 al & ar, 215, 231, David Hosking 100, Frans Lanting 17 al, 19 bl, 28, 32, 46, 49, 114 bl, 118 bcl, 142 bl, 152, 202, 243, Fritz Polking 132-3, 159, 193 bl, Gerry Ellis/Minden Pictures 117, 137 b, 195 b, 247, Ingo Arndt/Minden Pictures 211 a, John Watkins 148, Jurgen & Christine Sohns 13, 34, 53 cb, 112-3, 137 a, 198-9, Konrad Wothe/Minden Pictures 146, 147 b, 154 r, 193 a, 203 b, 218, 232, 241 ar, 269, Malcolm Schuyl 18 ar, 35, Paul Hobson 110 cl & cr **Fotolia** Baloncici 143 left ac, NiDerLander 143 right bl, Paul Cowan 143 right ac, Simone van den Berg 143 right al, soundsnaps 68 background, Suprijono Suharjoto 143 right ar, Tomo Jesenicnik 143 right bc, Tracey Kimmeskamp 142 right cl, Valery Sibrikov 143 right bcr **Getty Images** AFP 260-1, Digital Vision 17 br, 115 br, 125 r, 143 bcr, Hulton Archive 80, 81, 96, Ian Sanderson 119r, Jeff Foott 248, John Giustina 2, Michael Nichols 136 b, Michael Politza/National Geographic 250-1, National Geographic/Frans Lanting 177 r, 178, National Geographic/Michael Nichols 281, National Geographic/Paul Zahi 277, Paula Bronstein/Liaison 273, 275, Per-Anders Pettersson 266, Popperfoto 69 bl, Roberto Schmidt/AFP 267, Shuji Kobayashi 119 cr, Theo Allofs 124r, Timothy Laman 154 l, 155 r, Topical Press Agency 69 al, Veer/Antonino Barbagallo 70 background **istockphoto.com** Chanyut Sribua-rawd 56 background, Henk Bentlage 50 background, Matt Tilghman 52 background, pixonaut 54 background **Jay Ullal** courtesy hf Ullman/Tandem Verlag, publishers of *Thinkers of the Jungle* 105 a, 110 bl & r, 111 bl & r, 123, 270-1, 283 **Picture Desk** Kobal Collection/RKO 82, Kobal Collection/ Universal 83 al, Kobal Collection/Warner Brothrs 83 ac **National Geographic Stock** Hugo van Lawick 88-9, 194 a, b & c, Cyril Ruoso/Minden Pictures 108, 165 a, Gerry Ellis/Minden Pictures 134 a, Michael Nichols 91, 188-9, 201 a, Tim Laman 221 **Nature PL** Anup Shah 10-11, 20-21, 30, 33, 155 l, 161, 183, 185 a, 197 ar, 197 al, 224, Bruce Davidson 37, 185 b, Karl Ammann 134 b, Nature Production 176 r, Nick Garbutt 16 b, 50 a, T J Rich 36, Xu Jian 57 b **Press Association Images** AP 95;
Photolibrary age-fotostock/Wanyee Wanyee 241 br, David Curl 143 cr, Fritz Polking 98-9, Michael Fogden 143 br, Nigel Pavitt 142 right ar, Oxford Scientific/Clive Bromhall 60, 206, Steve Bromhall 66 **Photoshot** Gerard Lacz 16 a, 56, imagebroker.net 142 c, Mark Bowler 26, 141 ar, NHPA 230, 237, 240 c, 241 al, NHPA/Andy Rouse 40, 156, 177 l, NHPA/Cede Prudente 150, NHPA/Martin Harvey 41, 54c, 118 bcr, 149 a, 176 l, 191, NHPA/Nick Garbutt 142 ac, NHPA/Stephen Robinson 119cl, 158, Photoresearchers 122, 167, 195 a, 201 b, 228-9, Robert Vos 6-7, Tom McHugh 136 a, Woodfall Wild 45, 140 a, Woodfall Wild/Joe McDonald 129 **Shutterstock** Ly Dinh Quoc Vu 142 left al, Jim Lopes 142 left br, Shi Yali 142 left ar **Steve Bloom** 16, 17 ca & b, 18 bl, 19 cr, 27, 38, 42, 47, 116, 118 bl, 125 l, 127, 140 & 142 background, 154, 165 b, 169, 173 br, 179, 193 br, 208-9, 211 b, 213, 214, 217, 240 background, 246 **Still Pictures** Biosphoto/Berndt Fischer 147 a, Biosphoto/Cyril Ruoso 149 b, 258, Biosphoto/J L Klein & M L Hubert 142 right al, Biosphoto/Michel Gunther 180-1, 236, 253, 256, blickwinkel/H Schmidbauer 257, Larissa Siebicke/Das Fotoarchiv 143 left ar **The Dian Fossey Gorilla Fund International** www.gorillafund.org 86 **TopFoto** 69 br, Rene Dahinden/Fortean 70 al **Willie Smits** courtesy hf Ullman/Tamden Verlag, 278-9, 282

Index